THE MUSICAL
From Broadway to Hollywood

THE MUSICAL: *From Broadway to Hollywood*
Copyright ©2022 by Michael B. Druxman. All Rights Reserved.

No part of this book may be reproduced in any form or by any means, electronic, mechanical, digital, photocopying or recording, except for the inclusion in a review, without permission in writing from the publisher.

This book is an independent work of research and commentary and is not sponsored, authorized or endorsed by, or otherwise affiliated with, any motion picture studio or production company affiliated with the films discussed herein. All uses of the name, image, and likeness of any individuals, and all copyrights and trademarks referenced in this book, are for editorial purposes and are pursuant of the Fair Use Doctrine.

The views and opinions of individuals quoted in this book do not necessarily reflect those of the author.

The promotional photographs and publicity materials reproduced herein are in the author's private collection (unless noted otherwise). These images date from the original release of the films and were released to media outlets for publicity purposes.
Cover Illustration by Oisin McGillion Hughes.

Published in the USA by
BearManor Media
1317 Edgewater Dr. #110
Orlando, FL 32804
www.BearManorMedia.com

Softcover Edition
ISBN-10: 000
ISBN-13: 978-1-62933-898-9

Printed in the United States of America

Also by MICHAEL B. DRUXMAN:

Paul Muni: His Life and His Films
Basil Rathbone: His Life and His Films
Make It Again, Sam: A Survey of Movie Remakes
Merv
Charlton Heston
One Good Film Deserves Another: A Survey of Movie Sequels

THE MUSICAL

From Broadway to Hollywood

Michael B. Druxman

South Brunswick and New York: A. S. Barnes and Company
London: Thomas Yoseloff Ltd

For Howard and Judy Keel

Contents

Preface
Acknowledgments
Introduction
1. *On the Town* — 27
2. *Annie Get Your Gun* — 33
3. *Gentlemen Prefer Bondes* — 38
4. *Kiss Me, Kate* — 43
5. *Oklahoma!* — 49
6. *Guys and Dolls* — 55
7. *Carousel* — 61
8. *Pal Joey* — 66
9. *Damn Yankees* — 70
10. *Porgy and Bess* — 74
11. *West Side Story* — 79
12. *The Music Man* — 86
13. *Gypsy* — 91
14. *Bye Bye Birdie* — 96
15. *My Fair Lady* — 101
16. *The Sound of Music* — 107
17. *Stop the World—I Want to Get Off* — 113
18. *Finian's Rainbow* — 119
19. *Oliver!* — 125
20. *Hello, Dolly!* — 131
21. *Paint Your Wagon* — 138
22. *Cabaret* — 143
23. *Man of La Mancha* — 148
24. *Jesus Christ Superstar* — 155
25. *A Little Night Music* — 161
    A Musical Gallery — 167
    Index of Titles — 200

Acknowledgments

Grateful acknowledgment is made to the many individuals and organizations who gave of their time and their knowledge, loaned films for viewing purposes, and/or helped gather stills in the preparation of this book:

George Abbott, Academy of Motion Picture Arts and Sciences, Richard Adler, Michael and Beverly Ansara, Audio Brandon Films, Budget Films, Kingsley Candler, Betty Garrett, Leric Goodman, John Green, Arthur Hiller, Howard Keel, KHJ TV, Christopher Lee, Mervyn LeRoy, William Ludwig, Gordon MacRae, Millicent Martin, Jack Miller, Gene Nelson, Brock Peters, Stanley Rubin, Bill Sargent, Sol C. Siegel, Solters and Roskin, Charles Strouse, Jule Styne, Norton Styne, Thunderbird Films, Shani Wallis, Paul Francis Webster, and Meredith Willson.

Introduction

Talking pictures and the movie musical were born together. When, in *The Jazz Singer* (1927), Al Jolson said, "You ain't heard nuthin' yet," he not only signaled the end of the silent film, but also gave audiences their first taste of an exciting genre that was possible only through this revolutionary new technical process.

Interestingly, this first "talkie" was both a musical, and a musical that had its origins—storywise, that is—on the Broadway stage. Samson Raphaelson's play, *The Jazz Singer,* had debuted in New York in 1925 starring George Jessel, before it was subsequently purchased by Warner Brothers. Its screen treatment was primitive and, even by 1927 standards, dull, but it featured a singing Jolson. That was enough. The public lined up at the box office, and studio heads knew they had to convert to sound.

The new "baby" demanded to be fed. Its insatiable diet was one of words. Movie producers bought up well-known stage plays—good and bad—and also revamped previously purchased theatrical properties that had had their dialogue eliminated for filming as silent productions.

Musical pictures started appearing on studio production schedules. Many of these were over produced vaudeville-like entertainments, such as *Paramount on Parade* (1930) or Warners' *The Show of Shows* (1929), that featured a studio's contract players doing individual musical turns.

But a good many of the early song-and-dance productions came from Broadway. Eddie Cantor repeated his stage success in *Whoopee* (1930) for Goldwyn; Jolson brought his *Big Boy* (1930) to Warner Brothers; and, at Paramount, the Marx Brothers did *Animal Crackers* (1930) and several other of their stage successes. *Good News* (1930), *Rio Rita* (1929), *Sally* (1929), *The Desert Song* (1929), *The Vagabond King* (1930), *Sunny* (1930), and *Show Boat* (1929) were other early movie musicals that had begun life in the theater.

Most of these pioneer musical efforts suffered from the same problem that plagued dramatic films of the period. Sound may have been altering the face of the movie business, but so many difficulties were inherent in its use (microphone placement, muffling of the noisy camera) that moving pictures became very static. With the camera photographing from a soundproof booth, movies took on the appearance of a noisily filmed stage play, with actors forced to give rigid performances as they struggled to direct their words toward the often awkwardly placed mikes.

1933 found the influx of two very influential talents into the world of film musicals. Over at Warner Brothers, Busby Berkeley choreographed a series of fascinating, bizarre numbers—featuring dozens of scantily-clad beauties—that were inserted into such otherwise forgettable musicals as *Forty-Second Street* and *Footlight Parade*. Berkeley's intricate camera-oriented dance patterns gave musicals an aura of spectacle that they had not previously achieved.

And, at RKO, Fred Astaire made *Flying Down to*

The Vagabond King (1930). Jeanette MacDonald and Dennis King starred in this early adaptation of the Rudolf Friml operetta.

Rio with Ginger Rogers. Astaire's warmth and stylish precision dancing endowed his films with a creative elegance that made the top composers like Gershwin, Kern, and Berlin eager to write material for him. In a series of successful formula pictures with Rogers and others, the entertainer introduced many songs that have retained their popularity to the present day.

With few exceptions, most film musical adaptations of the thirties and forties bore scant re semblance to their stage originals. Sometimes the plot was completely changed; on other occasions, all but the major hit tunes were dropped in favor of new material. Then, again, the picture might be made as a non-musical. Sometimes only the pre-sold title remained from the Broadway show.

Cole Porter was just one composer who had his stage musicals cut and reshaped on numerous occasions. When, for example, his 1932 show, *The Gay Divorce,* became *The Gay Divorcee* (1934),

an Astaire/Rogers vehicle at RKO, only the classic "Night and Day" was retained from the original score, while new songs, including the Oscar winning "The Continental," were composed by others. A few years later, Porter's 1944 production of *Mexican Hayride* became a 1948 movie starring Abbott and Costello.

George Gershwin also had other composers' work interpolated into his shows. Jerome Kern's magnificent song, "The Last Time I Saw Paris," won the Oscar when inserted into Metro's 1941 release, *Lady Be Good,* with Eleanor Powell and Robert Young. This film had been adapted from the 1924 Gershwin stage smash starring Fred and Adele Astaire. Only the title and a few Gershwin tunes were retained from the original.

Gershwin's *Rosalie* starring Miss Powell was made in 1937, but the 1928 Broadway production now had an entirely new score by Cole Porter that included "In the Still of the Night." Two more

The Gay Divorcee (1934). Fred Astaire, Ginger Rogers, and Erik Rhodes. Only one song—"Night and Day"—remained from Cole Porter's original 1932 score.

Mexican Hayride (1948). This 1944 Cole Porter musical became a vehicle for Abbott and Costello.

Gershwin shows, *Girl Crazy* and *Strike Up the Band* (both staged in 1930), were bought by Metro and, with new story lines, were utilized as Mickey Rooney / Judy Garland vehicles. The composer fared best in the 1943 *Girl Crazy,* which retained several of his songs, but in *Strike Up the Band* (1940), the entire score—save the title number—was by Roger Edens.

Rodgers and Hart's *A Connecticut Yankee* debuted back in 1927, yet wasn't filmed by Paramount until 1948. Sadly, the original score was missing, and star Bing Crosby crooned a new set of pleasing numbers by Jimmy Van Heusen and Johnny Burke. *Babes in Arms,* also by Rodgers and Hart, was done by MGM with about half of the composers' numbers intact. The remainder were written by producer Arthur Freed. As the first of the Rooney/Garland backstage pictures ("Hey, kids, let's put on a show!"), it was quite successful, inspiring several similar entertainments about the juvenile entrepreneurs.

One show that was treated quite nicely was the Jerome Kern/Oscar Hammerstein II classic, *Show Boat,* remade by Universal in 1936 with a cast that included Allan Jones, Irene Dunne, Charles Winninger, Helen Morgan, and Paul Robeson. Al though the story line was expanded from the stage rendition to progress the plot forward a generation, the composers' original score was retained, virtually in its entirety. Many devotees argue that this black and-white version is superior to the Technicolor one MGM produced in 1951, which was closer to the stage in plot and starred Kathryn Grayson, Ava Gardner, Howard Keel, and Joe E. Brown.

There were several practical reasons why movie producers radically altered the expensive stage properties they purchased for their studios. Often the plots for these 1930s entertainments were of a simple boy-meets-girl formula that was designed merely to bridge the gaps between songs. The story might be acceptable on stage, but on the big screen its flimsiness would have been all too apparent.

Some shows were purchased with the idea of starring a particular studio contractee in the project and, as a result, the story would have to be revamped to fit that performer's talents, e.g., the Rooney/Garland backstage musicals.

From a musical standpoint, some of the less popular numbers in an original score might be dropped in order to shorten a film's running time. But certainly the primary reason that studios made extensive score interpolations was financial. If, for example,

Show Boat (1936). Charles Winninger played "Captain Andy" in the second filming of this Jerome Kern/Oscar Hammerstein II classic.

Paramount had used all of Cole Porter's original score when filming *Anything Goes* in 1936, it would have owned no interest in the music. However, by having other tunesmiths write new songs, the studio retained the valuable publishing rights to the interpolated material, which would allow it to license the songs for years to come. Only a handful of Porter tunes remained in that picture, which was remade in 1956—also with a primarily non-Porter score. In that latter production, Jimmy Van Heusen and Sammy Cahn did the interpolation. Bing Crosby, incidentally, starred in both versions.

Composers like Porter may not have appreciated the maiming of their scores but, considering the huge fees they were paid for their use, few objected too strongly. It has been only in recent years that composers have insisted they be given more control over how their work is treated on film. According to Jule Styne: "Today, when one of my shows is sold to movies, part of the deal states that any interpolations must be done by myself. This was the case when we sold *Funny Girl.* Bob Merrill, the lyricist, and I wrote a special title song for the picture."

In some musical adaptations, the producers might interpolate songs for the same composer's other works. Such was the case with Cole Porter's *Can Can* (1960) and Rodgers and Hart's *Pal Joey* (q.v.). Both films dropped lesser songs from the original scores, in favor of some of those composers' more endearing pieces.

Thanks to the likes of Nelson Eddy and Jeanette MacDonald, operettas, such as *Rose Marie* (1935) and *New Moon* (1940), were popular movie fare during Hollywood's Golden Era. Then, in the fifties, this classier kind of musical was revived with efforts like a third version of *The Desert Song* (1952) at Warner Brothers starring Kathryn Grayson and Gordon MacRae; a second rendition of *The Vagabond King* (1956) at Paramount with Miss Grayson and Oreste; and, at Metro, productions of *The Student Prince* (1954) and *Rose Marie* (1954), both with Ann Blyth, and *The Merry Widow* (1952) starring Lana Turner.

In these later pictures, the producers wisely decided that a certain amount of updating was required to make the shows palatable for modern audiences. These changes did not occur so much in the operettas' settings or music as in the dialogue and lyrics. Three-time Oscar-winning lyricist Paul Francis Webster, whose tunes include such standards as "The Shadow of Your Smile" and "Love Is a Many Splendored Thing," was hired to "doctor" the three aforementioned MGM projects, which had scores by such masters as Romberg, Friml, and Lehar. "The problem in this kind of lyric writing," says Webster, "is that you have to make the songs contemporary enough so that they'll work *today*, but still retain about the same mood and feeling as was in the original. In other words, you take out the archaic, yet are careful not to change things so much that the people who love the work will resent what you've done. There are some cases, of course, in which you'll interpolate altogether new songs."

While doing *Rose Marie,* Webster had the opportunity to work directly with the show's composer, Rudolf Friml. "He wasn't too sensitive about lyric changes in general," Webster recalls, "except when it came to using the word 'kiss.' On that he was impossible. He was probably in his late seventies then and refused to work on songs with 'kiss' in them because, he claimed, since kissing transfered

Rose Marie **(1954). Ann Blyth and Howard Keel starred in this lavish color remake of the popular operetta, which was first presented back in 1924.**

germs, it was unhygenic. We had to bring in a ghost writer to finish the song."

After the end of World War II, and especially with the advent of *Oklahoma!* (q.v.) in 1943, the elements that made up stage musicals (songs, dances, story, and characterization) became more integrated. Characters were more three-dimensional, and plots were also better thought out. No longer was it so easy for a studio just to drop a song or replace it with any other melody. The new show tunes had a specific function in developing either characterization or story line. True, there were still plenty of changes made in newly purchased shows, but as Hollywood entered the fifties and beyond, these alterations became less frequent and, when they were made, more care was taken so that the interpolations and book changes were compatible with the original material. Not only did the composers insist that this course be taken, but moviegoers throughout the country, who had already stopped rushing out to see most original movie musicals, let it be known that if they were going to pay to see an adaptation of a Broadway smash, they wanted to see something as close to the original as possible. Indeed, with studios seeking ways to lure audiences away from their televisions, motion picture rights to a hit Broadway musical became an increasingly valuable commodity.

Still, there were occasional instances when a successful show's entire score was eliminated so that the movie could be filmed as a straight comedy or drama. Warner Brothers shot *Fanny* in 1961 sans the lovely Harold Rome score that had been such an intergal part of the 1954 stage musical starring Ezio Pinza. And, in 1963, United Artists ignored the songs from the 1960 theatrical hit, *Irma la Douce,* yet still produced a funny Billy Wilder comedy with Jack Lemmon and Shirley MacLaine.

Major studio musicals featuring all-black casts have been few and far between. Surely one of the best to date was *Cabin in the Sky,* produced in 1942 by MGM and starring Lena Horne, Ethel Waters, Rex Ingram, and Eddie "Rochester" Anderson. The picture kept much of the Vernon Duke/John Latouche/Ted Fetter score that had worked so well in the stage production; however several numbers by Harold Arlen and E.Y. Harburg, including the show-stopping "Happiness Is Just a Thing Called Joe," were interpolated.

Cabin in the Sky also marked the film directing debut of Vincente Minnelli, who, over the years, has been responsible for some of Hollywood's greatest musicals. A craftsman of most exquisite taste, Minnelli's highly acclaimed work includes *Meet Me in St. Louis* (1944), *The Pirate* (1948), *An American in Paris* (1951), *The Band Wagon* (1953),* and *Gigi* (1958). Strangely, although he has directed several pictures adapted from stage musicals *(Brigadoon, Kismet, Bells Are Ringing,* and *On a Clear Day You Can See Forever),* none of these well-made films have achieved the same pinnacle of excellence set by his originals.

Otto Preminger directed two musicals with black casts in the 1950s. *Carmen Jones* (1954) was an adaptation of Oscar Hammerstin II's updated reworking of the Bizet opera, which had played New York in 1943. Harry Belafonte and Dorothy Dandridge starred in the picture. The other Preminger effort was Gershwin's *Porgy and Bess* (q.v.).

More recently, Daniel Mann directed the memorable Kurt Weill/Maxwell Anderson musical drama, *Lost in the Stars* (1974), which had debuted on Broadway in 1949, and Sidney Lumet did *The Wiz* (1978), an all-black rendition of *The Wizard of Oz.* Unlike the 1975 hit rock stage musical, Lumet's film—a vehicle for singer Diana Ross—reshaped the story so that Frank Baum's classic fantasy was now set in Harlem instead of Kansas. Most reviewers were critical of the Universal/Motown release, citing "overbearing logistics" as a major fault.

Though the Daniel Mann and both Preminger musicals delivered some good performances and offered generally fine renditions of the Bizet, Gershwin, and Weill scores, as films all three were disappointments. The fault was apparently in their direction.

Mann and Preminger are both capable dramatic directors but, when it comes to the musical, neither seems to possess that special talent necessary to bring this type of entertainment to life. In their musical numbers—whether they be solo or ensemble—these two directors simply photographed the action in lifeless medium shots or, in the case of *Lost in the Stars,* hand-held close-ups. Clearly, neither possesses that particular sense of musical style in staging, cinematography, and editing that makes the audience a part of the numbers, rather than just observers. It's probably an innate gift that has been bestowed on very few—Minnelli,

---

*The Band Wagon had been a revue on Broadway.

Carmen Jones (1954). Dorothy Dandridge and Harry Belafonte. Otto Preminger directed this adaptation of the Bizet opera, which had been reworked by Oscar Hammerstein II back in 1943 for a 502 performance Broadway run.

Gene Kelly, Bob Fosse, George Sidney and the like—but it's one that can determine just how satisfying these movies will be.

David Swift's lack of musical style was apparent in his otherwise competent direction of *How to Succeed in Business without Really Trying* (1967). By using multiple cameras to capture his numbers, Swift may have cut shooting time, but again audience involvement was absent. "The trick," he explained to the *Los Angeles Times* during production, "is not to take your audience into the scene but deliver the scene to the audience." Swift was surely successful in attaining *his* goal, however what he wound up with was a disappointment to those who fondly recalled the stage production.

Other disappointments: Richard Lester's cute little editing tricks made his *A Funny Thing Happened on the Way to the Forum* (1966) a confusion, and Ken Russell's unorthodox approach to *The Boy Friend* (1971) resulted in a once charming musical diversion becoming a bore.

There are, of course, several dramatic directors who have functioned quite well in the musical medium. Some of the best musicals made have come from the likes of William Wyler, Robert Wise, and Carol Reed. Working with a good choreographer and musical director, these men and others have proven that they can make a number come to life before the camera. Their images involve their audiences. Both Wise and Reed won Oscars for their musical pictures, and Wyler, directing a musical for the first time, created in *Funny Girl* (1968) some of the most exciting sequences in memory, witness the helicopter tracking Streisand as she sang, "Don't Rain on My Parade."

Perhaps, then, the answer is not so much one of "musical sense" as of the difference between being a competent and a great *film* director.

The King and I (1956). Yul Brynner and Deborah Kerr. Brynner repeated the role he did on Broadway in 1951.

Proper casting . can be just as vital to a musical picture—or any movie, for that matter—as the choice of a director. It's difficult to imagine anyone but Yul Brynner playing the King in Rodgers and Hammerstein's *The King and I* (1956). He had done the part in the original 1951 Broadway production with Gertrude Lawrence, and subsequently won an Oscar for his portrayal in the film with Deborah Kerr. Similarly, nobody but Barbra Streisand would have been right for Fanny Brice in *Funny Girl,* nor could anyone else have delivered that same deft pixieish humor as Robert Morse in *How to Succeed in Business without Really Trying.* There is definitely a strong argument for, whenever possible, casting the original stage star in the movie.

Film actors who have taken over roles from the Broadway originators have, nevertheless, often proven themselves worthy. Debbie Reynolds's energetic performance was the best thing about the movie version of *The Unsinkable Molly Brown* (1964), and Doris Day, Shirley MacLaine, and Mitzi

***How to Succeed in Business Without Really Trying* (1966). Rudy Vallee and Robert Morse, repeating their original Broadway roles, sing "Grand Old Ivy."**

Gaynor were definite assets to *The Pajama Game* (1957), *Sweet Charity* (1968), and *South Pacific* (1958) respectively. These ladies were chosen over stage performers such as Mary Martin and Gwen Verdon because the producers believed they meant more to the *movie* box office than the original stars.

Other movie replacements have not been as happily cast. Topol was an adequate Tevya in Norman Jewison's adaptation of *Fiddler on the Roof* (1971), but he lacked the innate warmth that Zero Mostel and others had given the part on stage. And Lucille Ball had the zaniness but not the sophistication necessary to bring *Mame* (1974) to life.

Whenever any stage property comes to the screen, it is nearly always "opened up" to varying extents: Interior scenes are rewritten to play out-of-doors, sequences are added to dramatize what was only discussed on stage, new characters are added, extremely stylized stagings are toned down, and orchestrations are adapted for a larger sound than had been possible in the theater. In sum, everything is made "bigger" to accomodate the newer medium. Film is, after all, a much more realistic art form than the stage.

There are musical adaptations that have been quite successful on film *without* going through any major "opening up" process, but examples such as *The King and I, Oklahoma!,* and *My Fair Lady* (q.v.) are few and far between. More often than not, the result is like the Tony Award-winning show about the signing of the Declaration of Independence, *1776* (1972)—a delight on stage but, because of its talky and interior-bound subject matter, often a chore to watch on the screen. *Brigadoon* (1955), directed by Minnelli and starring Gene Kelly, was, because of budget considerations, shot on a soundstage. The result was certainly entertaining, yet had exteriors rather than cycloramas been used to simulate the Scottish Highlands, this work—with a lilting score by Lerner and Loewe—might have been better received than it was.

Sweet Charity on film was definitely enhanced by director Bob Fosse's creative touches that went far beyond what he'd done in the 1966 stage version. And *Camelot* (1967) may have had script difficulties, but its real exteriors mixed with plush sound stage exteriors and interiors helped create just the right combination of realism and fantasy necessary to tell this charming story of King Arthur.

Then there is the problem of opening up a play *too* much. Director Joshua Logan's color camera filters were annoying in his production of *South Pacific,* and, in the adaptation of Rodgers and Hammerstein's *Flower Drum Song* (1961), producer Ross Hunter gave us a far too opulent production, burying the magic that had kept theater audiences entertained for 600 performances

"A bad musical film is one that sticks to the play," Norman Jewison told the *Hollywood Reporter* in 1971. "With *Fiddler on the Roof,* the first thing I did was forget the play.

"Film borrows from the theatre and literature, but it's a bonafide art form in itself and must be approached from a different aspect. As for musicals, the old ones are fantasies, such as *The Wizard of Oz* or a Fred Astaire movie, but those won't work today because they aren't real. I had no intention in *Fiddler* of having those daughters sing 'Matchmaker' wile dancing with brooms as they did on the stage, or have villagers swinging out of doors on 'Tradition.'"

(Jewison's argument that the "integrated" musicals of yesterday—such as those with Astaire and Gene Kelly—are now dead is difficult to accept. The idea of an actor suddenly breaking into a song-and-dance in the middle of a conversation is an unreality that audiences have always accepted, even in the most realistic of musicals. It's part of the genre and, despite attempts by filmmakers to go in another direction, it will survive. Audiences who love musicals want that kind of escapism.)

At this writing, *Fiddler on the Roof* is ranked third only to *Grease* (1978) and *The Sound of Music* (q.v.) as the most successful movie musical that has come from the stage. Its domestic film rentals total over $34 million.

There is much to the old adage, "You can't argue with success." Nevertheless, a great many people who saw both the stage and screen versions of *Fiddler* have expressed the opinion that the work lost much in its translation to film.

The major criticism seems to be that, in his quest for realism, Jewison went too far. In the theater, for example, a great deal of the violence and persecution aimed at the Jewish people occurred offstage. The audience heard about these terrible incidents but did not see them. Also the small village of Anatevka was presented with a very sparse, stylized set.

Jewison, on the other hand, let audiences see—in starkly realistic terms—the village, a cold, damp and depressing location. Russian soldiers rode through

Fiddler on the Roof (1971). Israeli actor Chaim Topol was producer/director Norman Jewison's choice for the role of "Tevya."

the muddy streets, threatening the Jewish populace and, in one unnecessary sequence set in another town, were shown riding down a group of protesters. Realistic it most certainly was, but at the sacrifice of much of the play's vital intimacy and glow.

The realistic musical, despite its hazards, is definitely here to stay. One need only look at Fosse's *Cabaret* (q.v.) to see that this approach can work very well indeed.

An interesting phenomenon we're seeing more of these days is the original movie musical that is later adapted to the stage. Warner Brothers' 1953 musical of *Calamity Jane,* starring Doris Day and Howard Keel, featured a bright score by Paul Francis Webster and Sammy Fain. Even though one tune, "Secret Love," won the Academy Award, the picture was a financial disappointment. A few years later the composers added a few more numbers, and the *stage* version of *Calamity Jane* had its formal debut at the St. Louis Municipal Opera with Carol Burnett in the lead. The comedienne later did the show as a television special. Martha Raye and Ginger Rogers are others who have done the musical on the dinner theater *I* summer stock circuit.

The composers, at this point, have no interest in a New York production. "We're going to play the show as long as we can around the country before we take it to Broadway," says Webster. "It's been playing successfully for years and we figure that if we got bad New York reviews, that might kill further interest in it."

Gigi and *Seven Brides for Seven Brothers* (1954) are two other musicals that have gone the reverse route—from movies to the stage.

The current trend in the movie musical appears to be toward rock. The success of *Jesus Christ Superstar* (q.v.) and *Tommy* (1975) inspired more ventures into this exciting area of popular music, with adaptations of such shows as *Grease* and *The Wiz,* both appearing in 1978.

Musicals are enormously expensive to produce on film. The studio investment in *Camelot,* for example, was over twelve million dollars, and the budget for *Hair* (1979), ten million. Thus it's not surprising that these entertainments are made so infrequently nowadays. There are, in fact, many stage successes that have yet to be made into pictures. *The Most Happy Fella, Fiorello, Redhead, Wildcat, Pipe Dream, No Strings, Follies, Company, I Do, I Do, The Fantasticks, Little Me,* and *House of Flowers* are just a dozen in a long list.

Two recent blockbusters—A *Chorus Line* and *Annie*—have been purchased by Universal and Columbia respectively, and their forthcoming productions assure us that the genre remains alive and well.

Calamity Jane (1953). **Doris Day and Howard Keel. This original Warner Brothers musical was later turned into a stage play.**

Introduction to the 2022 Reprint Edition

Not a lot has changed with the movie musical since the original edition of this book was published in 1980.

At the end of the Introduction of that first book, I listed a dozen hit Broadway musicals that had yet to be produced as a motion picture.

Brad Sullivan and Joel Grey played the feuding fathers in *The Fantasticks* (2000).

To date, only one of those musicals (*The Fantasticks* in 2000) has made it to the screen, and that picture did not do well at the box-office.

Still, there have been quite a few other Broadway musicals adapted to film since 1980. Some were very good, while others…

The Fantasticks was a troubled film production from the start.

The original off-Broadway production opened in New York at the Sullivan Street Playhouse on May 3, 1960, ran for 17,162 performances, closing in 2002.

That cast included Jerry Orbach (TV's *Law & Order*).

Songs included "*Try to Remember,*" "*There Were You*" and "*Round and 'Round*".

The plot deals with two neighboring fathers who trick their children into falling in love by not only pretending to feud, but also by arranging for a professional bandit to stage a fake kidnapping.

Paramount Pictures owned the rights to the property in the late 1960s.

Originally, Gower Champion was set to direct with Howard W. Koch producing. During the summer of 1972 Paramount even paid for a trip to Italy for Champion, and songwriters Harvey Schmidt & Tom Jones, to scout locations.

By January 1973, for whatever reason, the picture had been called off.

The movie, produced and released by MGM/UA, and featuring Joel Grey in the cast, was, finally, filmed in 1995… then shelved for five years.

The released version was re-edited by Francis Ford Coppola with the consent of director Michael Ritchie.

Running time was 110 minutes before being whittled down to 85 minutes for theatrical release.

Edmund Guthmann, in the *San Francisco Chronicle*, said of the film: "The Fantasticks *has slow patches and requires a generous suspension of disbelief. But it's also sweet and optimistic -- a welcome antidote to gloom.*".

For some inexplicable reason, "*Try to Remember,*" the hit song from *The Fantasticks*, as well as two other numbers, were eliminated from the theatrical release. However, they were restored for the DVD version.

When my original book went to press, two other Broadway musicals had been announced for filming, but had yet to go into production.

They were *Annie* and *A Chorus Line*.

Aileen Quinn played the title role in the film version of *Annie* (1982), and Carol Burnett was "Miss Hannigan".

Annie, based on Harold Gray's "Little Orphan Annie" comic strip, made it to the screen in 1982.

The original Broadway production had opened in 1977 and ran for nearly six years. It had music by Charles Strouse, lyrics by Martin Charnin, and a book by Thomas Meehan.

It won seven Tony Awards, including "Best Musical," and featured the hit song, "*Tomorrow*".

John Huston (*The Maltese Falcon*, *Treasure of Sierra Madre*), for some curious reason, was chosen to direct the film version of the hit musical.

His cast included Albert Finney as "Daddy Warbucks," Carol Burnett as "Miss Hannigan," Ann Reinking as "Grace Farrell," Tim Curry as "Rooster," and Aileen Quinn as "Annie".

Auditions for the title role spanned two years, 22 cities, 8,000 interviews, and 70 actresses. Drew Barrymore, in fact, auditioned for the title role.

The stage version of *Annie* ends at Christmas. The movie changed it to the 4th of July, because it was shot during the summer, and getting enough fake snow to cover the grounds of the New York mansion was far too expensive.

The songs "Dumb Dog," "Sandy," "Let's Go to The Movies," "We Got Annie," and "Sign" were written expressly for the movie. Several other songs from the Broadway musical were dropped.

Set on the bare stage of a Broadway theater, *A Chorus Line* has music by Marvin Hamlisch, lyrics by Edward Kleban, and a book by James Kirkwood Jr. and Nicholas Dante.

Despite being the tenth biggest grossing movie of 1982, *Annie* didn't make a profit because of its exorbitant production costs.

Roger Ebert, giving the film a rating of *3-Stars*, said in his review: "*It's like some kind of dumb toy that doesn't do anything or go anywhere, but it is fun to watch as it spins mindlessly around and around.*"

The cast of the 1986 filming of *A Chorus Line*.

Seventeen Broadway dancers audition for spots in a chorus line. We get a glimpse into the personalities of each of the performers and the choreographer, as they describe the events that have shaped their lives and their decisions to become dancers.

Opening in 1975, and directed by Michael Bennett, the original Broadway production ran for 6,137 performances, becoming the longest-running production in Broadway history until surpassed by *Cats* in 1997.

The play contains many excellent songs, including "What I Did for Love," "At the Ballet," "I Can Do That," and the show-stopper, "One".

The 1986 movie version was directed by Richard Attenborough, and featured Michael Douglas as the play's director/choreographer. The rest of the cast was, for the most part, unknown.

A Chorus Line belongs on the stage.

It should never have been adapted to film, because, on film, it becomes bigger. And, with a star like Michael Douglas headlining the cast, his role has been enlarged out of proportion to its function in the play.

In the stage production, the role of "Cassie" (played by Alyson Reed in the film) is really the star part. Donna McKechnie played her on Broadway, and the director (Douglas in the film) is hardly ever seen.

Paul Attanaslo in the *Washington Post* said: "*A Chorus Line is a kind of* "Murphy's Law: The Motion Picture" -- *everything that can go wrong does. Based on the long-running Broadway musical, it's not an adaptation so much as an assassination, a case study in how not to bring a play to the screen.*"

"Audrey II," the homicidal plant in *Little Shop of Horrors* (1986).

Little Shop of Horrors (1986) began life in 1960, as a low budget film, directed by Roger Corman. It tells of a geeky florist shop worker who finds out his Venus flytrap has an appetite for human blood.

In 1982, the film's plot served as the basis for a horror comedy rock musical, with music by Alan Menken and lyrics and a book by Howard Ashman.

The musical premiered Off-Off-Broadway before moving to the Orpheum Theatre Off-Broadway, where it had a five-year run. It later received numerous productions in the U.S. and abroad, and a subsequent Broadway production.

Aside from the title song, the Menlen/Ashman score includes "Skid Row (Downtown)," "Somewhere That's Green," "Suddenly, Seymour," and "Mean Green Mother from Outer Space,".

The 1986 film, directed by Frank Oz, stars Rick Moranis (as the geeky flower shop worker), Ellen Greene, Vincent Gardenia, Steve Martin, and Levi Stubbs as the voice of "Audrey II," the homicidal plant. The film also featured special appearances by Jim Belushi, John Candy, Christopher Guest and Bill Murray.

After a delay needed for the studio to complete a revised, happier ending, *Little Shop of Horrors* was released at the end of 1986.

Janet Maslin in The New York Times, called the film, "*a full-blown movie musical, and quite a winning one.*"

The film would, later, become a major hit on home video.

Adapted from the play by Larry L. King and Peter Masterson, *The Best Little Whorehouse in Texas* (1982) is not really a very good picture.

Charles During played the Governor of Texas in *The Best Little Whorehouse in Texas*, and not only "stopped the show" with his song, "Sidestep," but also earned an Oscar nomination.

With original music and lyrics by Carol Hall, it is based on a story by King that was inspired by the real-life Chicken Ranch in La Grange, Texas.

The Best Little Whorehouse in Texas opened on Broadway at the 46th Street Theatre on June 19, 1978, and ran for 1,584 performances.

Directed by Colin Higgins, the film adaptation of the musical starred Burt Reynolds and Dolly Parton, but the stand-out performance in the picture was delivered by Charles Durning, who played the Governor of Texas, and delivered the show-stopping number, "Sidestep". He received a Supporting Actor Oscar nomination for his performance.

Of the overall film, Janet Maslin in *The New York Times*, said: "*It's just plain dull.*".

In 1983, Universal Pictures filmed Gilbert and Sullivan's operetta, *The Pirates of Penzance*. It starred Kevin Kline as "The Pirate King," Angela Lansbury and Linda Ronstadt. Wilford Leach directed.

Kline had previously won the Tony Award in 1981-82 for a production of the play in New York.

Of the film, Bob Thomas of Associated *Press* said: "*The singing is full and rich, the casting inspired.*"

The movie's failure at the box-office had nothing to do with the reviews, which were, for the most part, positive. The real problem lay with Universal Pictures' decision to release this movie simultaneously to *Select TV* and to theaters.

Theater owners were so upset, that they boycotted this movie. In the end, a grand total of ninety-two theaters agreed to show it. It did enjoy a long run only at a Washington, D. C. theater, where it played several weeks.

Madonna played the title role in the film version of *Evita* (1996).

Certainly the most memorable musical to emerge from the 1990s was *Evita*, with music by Andrew Lloyd Webber and lyrics by Tim Rice. It dealt with the life of Argentine political leader Eva Perón, the second wife of Argentine president Juan Perón.

The project began as a 1976 rock opera concert album, eventually becoming a stage play two years later. After its London debut, the musical moved to Broadway where it played for 1,567 performances, and received the Tony Award for Best Musical.

In the New York production, Patti LuPone played the title role, Mandy Patinkin was Che Guevara, and Bob Gutton was Perón.

The song, "*Don't Cry for Me Argentina,*" became an instant standard.

Ken Russell was the first director attached to the film. His initial choice to play Eva was Barbra Streisand, who turned him down. Meryl Streep, Liza Minnelli and Charo were also considered for the role.

Eventually, Alan Parker directed the 1996 film version of *Evita*, which featured Madonna in the title role, Antonio Banderas as Che, and Jonathan Pryce as Perón.

In the *San Francisco Chronicle*, Octavio Roca wrote: "*Evita was never meant to be a lesson in Argentine history, but it became one of the great musicals of all time anyway. If it is not one of the great movie musicals, it is close enough.*"

Quite a few Broadway musicals were adapted to film in the first decade of the 2000s.

Certainly the most revered was *Chicago* (2002), which won six Academy Awards, including Best Supporting Actress (Catherine Zeta-Jones) and Best Picture.

Renee Zellweger and Richard Gere in the "We Both Reached for the Gun" number from the 2002 filming of *Chicago*.

The musical was based on a 1942 Ginger Rogers film, *Roxie Hart*, which was, in turn, a remake of a silent picture, *Chicago* (1927) with Phyllis Haver.

Chicago was, in fact, inspired by a true story.

The characters of "Roxie Hart" and "Velma Kelly" (Renée Zellweger and Catherine Zeta-Jones in the musical film) were based by the real-life cases of two Chicago women -- Beaulah Annan and Belva Gaertner -- who each murdered their lovers in 1924. Both women were tried in separate cases, which became media sensations in Chicago, and both were acquitted.

Starring Gwen Verdon and Chita Rivera as "Roxie Hart" and "Velma Kelly," with Jerry Orbach as their attorney, "Billy Flynn," the original Broadway production opened in 1975 at the 46th Street Theatre and ran for 936 performances, until

1977. Bob Fosse directed and choreographed the original production.

Music was by John Kander, lyrics by Fred Ebb, and book by Ebb and Fosse. Songs include: "All That Jazz," "Razzle Dazzle," "Cell Block Tango," and "Roxie".

At this writing, the 1996 revival production holds the record as the longest-running *musical revival* and the longest-running American musical in Broadway history. It is the second longest-running show to ever run on Broadway, behind only *The Phantom of the Opera*.

When the movie rights to *Chicago* were originally bought by producer Martin Richards in the 1970s, Bob Fosse was to be involved with the project, and Goldie Hawn, Liza Minnelli, and Frank Sinatra were announced as the stars, but Fosse's death in 1987 ended that attempt at a movie version.

The 2002 film was shot in Toronto, Ontario, Canada, and directed by Rob Marshall.

In the play, the musical numbers were staged as vaudeville acts; the film respects this but presents them as cutaway scenes in the mind of the Roxie character, while scenes in "real life" are filmed with a harder-edge.

The movie did well at the box-office, and critical response was good. Joining Renée Zellweger and Catherine Zeta-Jones in the film were Richard Gene as "Billy Flynn," as well as Queen Latifah and John C. Reilly, who were both nominated for Oscars.

Writing in *Variety*, David Rooney said: "*A stylish cast and some clever scripting solutions help* Chicago *make the transition from stage to screen with considerable appeal intact. But despite these assets, plus the enduring kick of the superlative Kander & Ebb song score, this film version dilutes a good deal of the live show's sizzle and wit.*"

Apparently, the Motion Picture Academy didn't agree with Mr. Rooney.

Lon Chaney, Claude Rains and Herbert Lom had all played the title role in *The Phantom of the Opera* on the screen, but those films were certainly overshadowed when the classic horror tale was turned into a stage musical in 1986.

The Phantom of the Opera was based on a 1910 novel of the same name by Gaston Leroux. The story revolves around a beautiful soprano, Christine Daaé, who becomes the obsession of a mysterious, disfigured musical genius, who lives in the subterranean tunnels beneath the Paris Opéra House.

Gerard Butler and Emmy Rossum in *The Phantom of the Opera* (2004).

The musical featured music by Andrew Lloyd Webber, lyrics by Charles Hart, and a book by Richard Stilgoe. "The Music of the Night," "Think of Me," "Masquerade," and "The Phantom of the Opera" were standout songs in the play

Phantom won the 1986 Olivier Award and the 1988 Tony Award for Best Musical, and Michael Crawford (in the title role) won the Olivier and Tony Awards for Best Actor in a Musical. *Phantom* is currently the longest running show in Broadway history, and celebrated its 10,000th Broadway performance in February of 2012, the first production ever to do so. It is the second longest-running West End musical, after *Les Misérables*.

After Hugh Jackman was forced to drop out of the motion picture version because of other commitments, Gerard Butler played the title role in the 2004 Warner Bros. filming of the musical, which was directed by Joel Schumacher at London's Pinewood Studios. The cast also included Emmy Rossum as Christine.

Prior to the making of the film, Gerard Butler had never had a proper singing lesson, so when he was recording "Music of the Night," he said, "it was quite difficult, considering how long you have to hold the ending note."

Derek Elley in *Variety* said: "Sumptuous pic version, which evokes the original show while working as a movie in its own right, is lit by a radiant, vocally lustrous performance by teenaged Emmy Rossum as the Phantom's muse and has a widescreen sweep and musical fluidity that avoids enervating, 'Moulin Rouge!'-like flashiness.".

On the other hand, Steve Davis of the *Austin Chronicle* echoed a feeling that I had about the score after I listened to the original cast recording: *"Is it just me, or are there only three melodies in this musical, each recycled over and over, ad nauseam?"*

In 2005, a pair of hit musicals appeared on cinema screens.

Nathan Lane and Matthew Broderick in *The Producers* (2005)

Mel Brooks' *The Producers* was adapted from his 1967 movie of the same name. That classic comedy, written and directed by Brooks, had starred Zero Mostel and Gene Wilder as two stage-play producers who devise a plan to make money by producing a sure-fire flop: a musical about Adolph Hitler.

Unfortunately, their illegal scheme to cheat their investors backfires when their play becomes an enormous hit.

Working with Thomas Meehan on the book, Brooks' musical adaptation of *The Producers*, opened at the St. James Theater in New York in April, 2001. It starred Nathan Lane and Matthew Broderick in the roles played by Mostel and Wilder in the original film. It ran for 2,502 performances, winning a record-breaking twelve Tonys, including the 2001 Tony Awards for the best musical, book and score.

Among the songs in the musical were: "I Wanna Be a Producer," "We Can Do It," "Keep it Gay," "That Face," "Prisoners of Love," and, of course, "Springtime for Hitler," which was also in the original film.

The 2005 motion picture rendering of the musical was directed by Susan Stroman, director of the Broadway production, and featured both Lane and Broderick in the roles that they had originated on stage. Also in the cast were Uma Thurman, Will Ferrell and Gary Beach.

Writing in *Variety*, Todd McCarthy said: "The Producers *comes full circle back to the screen 38 years after the original with much, if not all, of its mirth intact. Reproducing Mel Brooks' still-running Broadway smash so literally you can practically see the proscenium arch, new pic is undeniably stagy, even clunky, and its commercial fate rests on whether auds find the blunt theatrical artifice and playing-to-the-balcony performance style off-putting or endearing.".*

The Producers garnered, generally, mixed reviews from critics and was a commercial failure, earning $38 million worldwide from a $45 million budget.

Rent is a rock musical, with music, lyrics, and book by Jonathan Larson.

It is loosely based on Giacomo Puccini's 1896 opera, *La Bohème*.

It tells the story of a group of impoverished young artists, struggling to survive and create a life in Lower Manhattan's East Village, under the shadow of HIV/AIDS.

The musical was first seen in a New York workshop in 1993.

It moved to Broadway's larger Nederlander Theatre in April of 1996.

On Broadway, *Rent* won several awards, including the Pulitzer Prize for Drama and the Tony Award for Best Musical.

Tragically, Larson, the show's creator, was not there to share in his play's acclaim. He died, suddenly, the night before the January, 1996, off-Broadway premiere.

Aside from the title number, among the songs in the play are "One Song Glory," "Season of Love," and "Goodbye, Love".

Directed by Chris Columbus, the 2005 Columbia Pictures movie starred Anthony Rapp, later of "Star Trek: Discovery" fame.

Rent is not a bad movie.

It's just a depressing one.

And, that's all I really care to say about it.

In 2006, Jamie Fox, Beyonce, Eddie Murphy and Jennifer Hudson, who won an Oscar for her supporting performance, appeared in the film version of *Dreamgirls*. Although the writer (Tom Eyen) denied it, the musical, was, in all likelihood, based on the story of Diana Ross and "The Supremes".

Anthony Rapp in *Rent* (2005)

Henry Krieger wrote the music, and lyrics and book were by Eyen.

Jennifer Hudson won an Oscar for her performance in *Dreamgirls* (2006).

The musical follows the story of a young female singing trio from Chicago, called "The Dreams," who become music superstars.

Starring Jennifer Holliday, *Dreamgirls* opened in December, 1981, at the Imperial Theatre on Broadway. The play won six Tony Awards.

Bill Condon directed the film adaptation of *Dreamgirls*.

In the *Austin Chronicle*, Steve Davis called the movie: *"an infectious experience of sequins and songs that lives up to the hype."*.

Interestingly, although the film earned eight Oscar nominations, it was not nominated for Best Picture.

With music and lyrics by Marc Shaiman and Scott Wittman and a book by Mark O'Donnell and Thomas Meehan, *Hairspray* (2002) was based on John Waters' 1988 film of the same name.

The plot concerns plump teenager Tracy Turnblad, who teaches 1962 Baltimore a thing or two about integration after landing a spot on a local TV dance show.

The gimmick of the staging is that Tracy's mother is portrayed by an actor "in drag". In the 1988 film, the part was played by Divine, with Jerry Stiller cast as "her" husband. In the 2007 version, John Travolta was the mother, with Christopher Walken as the man in "her" life.

The musical opened in Seattle in 2002, and moved to Broadway later that year. The original Broadway cast included Marissa Jaret Winokur and Harvey Fierstein in the lead roles of Tracy and Edna, her mother.

In 2003, *Hairspray* won eight Tony Awards, including one for Best Musical, out of thirteen nominations. It ran for 2,642 performances, and closed in January of 2009.

Nikki Blonsky played Tracy in the 2007 film adaptation of the musical, which was directed by Adam Shankman.

Among the songs retained for the movie were: "Good Morning Baltimore," "You Can't Stop the Beat," "Without Love," and "I Know Where I've Been".

According to press releases, it took John Travolta four hours each day to put on the 30-pound fat suit and 5 gel-filled silicone face prosthetics to become Edna Turnblad. Aside from an entertaining performance, his efforts earned him a Golden Globes nomination.

Betty Jo Tucker, writing in *Reel Talk*, said: *"Just like* Singin' in the Rain, Hairspray *moves along without a single dull moment from beginning to end. And it features the same type of energetic performances as well as musical numbers that are uninterrupted by cut-away shots, so viewers can enjoy the talents displayed by actors, musicians, composers (Marc Shaiman and Scott Wittman) and choreographer extraordinaire Adam Shankman, who also served as director."*

John Travolta in *Hairspray* (2007)

Hairspray earned over two hundred million dollars at the American box-office.

2008 gave us the film version of *Mama Mia*, a "jukebox musical" which featured the songs of ABBA. Meryl Streep, Amanda Seyfried, Colin Firth, Julie Walters, Christine Baranski, and Pierce Brosnan were among those who appeared in the movie.

After playing in several U.S. cities, the staging had opened in New York in 2001.

In 2009, there were films adaptations of two more stage musicals, *Fame* and *Nine*, both inspired by previously produced movies.

And, in 2011, we had *Footloose*, which also had its start on film before it found its way to the stage.

However, in this writer's opinion, the next *major* Broadway musical to become a film was *Les Miserables (aka: Les Miz)*, adapted from Victor Hugo's classic novel.

Hugo's book is the story of Jean Valjean, a French peasant, who is released from prison after serving nineteen years for stealing a loaf of bread for his sister's starving child. Breaking parole, yet still leading an honest life, he is pursued for years by Javert, a fanatical police officer determined to bring him to "justice".

Les Miserables has been filmed many times as a straight drama, most memorably in 1935. That production was directed by Richard Boleslawski, and featured Fredric March as Jean Valjean and Charles Laughton as Javert.

The sung-through musical adaptation of Hugo's work was by Claude-Michel Schönberg (music), Alain Boublil and Jean-Marc Natel (original French lyrics), and Herbert Kretzmer (English lyrics).

Premiering in Paris in 1980, the original French musical had direction by Robert Hossein. Its English-language adaptation by producer Cameron Mackintosh has been running in London since October 1985, making it the longest-running musical in the West End and the second longest-running musical in the world after the original Off-Broadway run of *The Fantasticks*.

The Broadway production, featuring Colin Wilkinson as "Jean Valjean" and Terrence Mann as "Javert", opened March in March of 1987 and ran until May of 2003, closing after 6,680 performances. The show was nominated for 12 Tony Awards, at that time a record number, of which it won eight, including Best Musical and Best Original Score.

Among the more memorable songs in *Les Miz* are: "I Dreamed a Dream," "Suddenly," "On My Own," "One More Day" and "Do You Hear the People Sing?".

The 2012 film version of the musical was directed by Tom Hooper, and featured Hugh Jackman as "Jean Valjean" and Russell Crowe as "Javert".

Interestingly, the film's vocals were recorded live on set using live piano accompaniments played through earpieces as a guide, with the orchestral accompaniment recorded in post-production, rather than the traditional method where the film's musical soundtracks are usually pre-recorded and played back on set to which actors lip-sync.

Writing in the *The Guardian*, Peter Bradshaw said: *"Even as a non-believer in this kind of 'sung-through' musical, I was battered into submission by this mesmeric and sometimes compelling film ..."*.

Nominated in eight categories, *Les Miz* won three Oscars, including one for Anne Hathaway (Best Supporting Actress), who played "Fantine," and sang "I Dreamed a Dream".

Rock of Ages was another "Jukebox musical," this one built around classic rock songs from the

1980s. The play featured songs from Styx, Journey, Bon Jovi, Pat Benatar, Twisted Sister, Steve Perry, Poison and Europe, among other well-known rock bands. It opened on Broadway in April of 2009.

Hugh Jackman played "Jean Valjean: in *Les Miserables* (2012).

Adam Shankman directed the 2012 film version, which features Tom Cruise, Alec Baldwin, Paul Giamatti, Catherine Zeta-Jones and Russell Brand.

The plot deals with a small-town girl (Julianne Hough) and a city boy (Diego Boneta) who meet on the Sunset Strip while pursuing their Hollywood dreams.

Writing in *The Hollywood News*, David Bennett said: "If you're craving some nostalgia, by all means tie a bandana around your bicep, thrust your rock horns aloft and you'll enjoy some laughs along the way too. If not, be warned – you'll leave with a headache and a horribly sugary taste in your mouth."

In 2014, two more stage musicals found their way to the motion picture screen.

Into the Woods had music and lyrics by Stephen Sondheim and book by James Lapine. The book intertwines the plots of several Brothers Grimm fairy tales (e.g. *Little Red Riding Hood, Jack and the Beanstalk, Rapunzel, Cinderella*), exploring the consequences of the characters' wishes and quests. Indeed, it is a much darker view than when our parents read us these stories when we were children.

The musical premiered on Broadway in November 5 of 1987, and won several Tony Awards, including Best Score, Best Book, and Best Actress in a Musical (Joanna Gleason).

Certainly the most memorable songs in the play were the title song, "Into the Woods," and "Children Will Listen".

Little Red Riding Hood (Lilla Crawford) in *Into The Woods* (2014)

The 2014 movie adaptation, from Walt Disney Productions, was directed by Rob Marshall. It starred Meryl Streep, Emily Blunt, James Corden, Anna Kendrick, Chris Pine, Tracey Ullman, Christine Baranski and Johnny Depp.

Writing for *Slant Magazine*, R. Kurt Osenlund said of *Into The Woods*: "*Even before the film's climax...Into the Woods—which runs an exhausting 124 minutes—already starts to unravel faster than Rapunzel's severed braid. Plotlines run rampant, interest wanes, and the brazen morals so saucily delivered in the movie's first two thirds become buried in literal and figurative final-act rubble.*"

Nevertheless, the film grossed over $213 million worldwide, and received three Academy Awards nominations.

Jersey Boys tells the story of four young men, led by Frankie Valli, from the wrong side of the tracks in New Jersey, who formed the 1960s rock group, The Four Seasons.

Written by Marshall Brickman and Rick Elice, the jukebox musical is narrated by each member of the band who gives his own perspective on its history and music.

Songs include "Big Girls Don't Cry", "Sherry", "December 1963 (Oh, What A Night)", "My Eyes Adored You", "Can't Take My Eyes Off You", "Who Loves You", and "Rag Doll".

Jersey Boys ran on Broadway from 2005 to 2017, and won four Tony Awards, including Best Musical.

Clint Eastwood directed the 2014 film adaptation of *Jersey Boys*.

Despite being pressured to cast more famous leads, Eastwood refused, stating, "*You've got people who've done 1,200 performances; how much better can you know a character?"

John Lloyd Young played "Frankie Valli" in Clint Eastwood's filming of *Jersey Boys* (2014).

Thus, John Lloyd Young played "Frankie Valli" in the film and "Erich Bergen" was "Bob Gaudio". Both actors were from the stage production.

Also cast was in the movie was Michael Lomenda as "Nick Massi," and Vincent Piazza, played "Tommy DeVito". Indeed, the only recognizable name in the cast are Christopher Walken, as a Mob boss.

Writing for Roger Ebert, Odie Henderson said: "*A movie as grungy, profane and blue-collar as* Jersey Boys *needs to feel more alive. When the music's not playing,* Jersey Boys *starts to lull you into mild lethargy.*"

Andrew Lloyd Webber's *Cats* (1981), a sung-through musical, had been filmed as a direct-to-video production in 1998. That cast included Elaine Page and John Mills. Directed by David Mallet, the production had been restaged on a new set, and was not filmed with an audience.

Cats was based on *Old Possum's Book of Practical Cats* (1939) by T. S. Eliot. The play tells of a group of cats, and the night they decide which cat will ascend to the "Heaviside Layer", and come back to a new life.

The musical features the hit song, "Memory".

Cats opened to positive reviews at the New London Theatre in the West End in *1981, running for twenty-one years,* and then to mixed reviews at the Winter Garden Theatre on Broadway in 1982, running for eighteen years. Initially, Betty Buckley played "Grizebella," the Glamour Cat, in that production, winning a Tony Award for her performance.

In fact, *Cats* won numerous awards including Best Musical at both the Laurence Olivier and Tony Awards.

In 2019, Tom Hooper, in his second film after *Les Miserables*, directed a new production of *Cats*, this one for theatrical release.

Judi Dench played "Old Deuteronomy" in *Cats* (2019)

The film, released by Universal, features an ensemble cast, including James Corden, Judi Dench, Jason Derulo, Idris Elba, Jennifer Hudson, Ian McKellen, Taylor Swift, Rebel Wilson, and Francesca Hayward.

Reviews were not good. Critics did not like the editing, nor the visual effects. The film was a box-office bomb.

Writing in *Variety*, Peter Debruge called *Cats*: "*one of those once-in-a-blue-moon embarrassments that mars the résumés of great actors... and trips up the careers of promising newcomers.*"

At this writing, two more musicals adapted from Broadway plays are scheduled to be released before the end of 2021.

In the Heights is a feature version of the Broadway musical, in which a bodega owner has mixed feelings about closing his store and retiring to the Dominican Republic after inheriting his grandmother's fortune.

After a 2007 off-Broadway run, the show opened on Broadway in March 2008. It was nominated for thirteen Tony Awards and won four, including Best Musical.

The film is adapted from the play by Quiara Alegria Hudes and Lin-Manuel Miranda. It is directed by Jon M. Chu, and features Jimmy Smits in the cast.

Also set for a 2021 release is Steven Spielberg's remake of *West Side Story*.

That Leonard Bernstein/Stephen Sondheim/Arthur Laurent musical debuted on Broadway in 1957, and was later adapted into a motion picture in 1961.

Co-directed by Jerome Robbins and Robert Wise, that film, which starred Natalie Wood, Richard Beymer and Russ Tamblyn, swept the Academy Awards; winning ten Oscars, including Best Picture, Director, Supporting Actress (Rita Moreno) and Supporting Actor (George Chakaris).

[*For further information on that production, see the chapter on West Side Story in this book.*]

Why would Stephen Spielberg remake a film classic?

Aside from the fact that all of the character in his version will be played by virtually unknown Hispanic actors, I have no clue.

Apparently Rita Moreno likes the idea. She is also in the cast, but not playing to role of "Anita," which she did in the original.

She's "Valentina," which is a re-imagining of the character of Doc (*Actor Ned Glass in the original*), the store owner who saved Moreno's character in the 1961 film from nearly being raped in his drugstore.

With Spielberg at the helm, I'm sure that the new *West Side Story* will be a good film.

But then, I come back to my original question:

Why would Stephen Spielberg remake a film classic…when there are so many other superb Broadway musicals out there that are waiting to be adapted to film?

THE MUSICAL
From Broadway to Hollywood

1
On the Town
(1949)

Metro-Goldwyn-Mayer's *On the Town* is a film that has not worn as well as some musicals produced earlier by that studio. It contains no truly memorable songs, its plot about three sailors on leave has been done many times before and since, and its once fresh staging of musical numbers, compared to what has been done in subsequent movies, is now somewhat primitive. Yet, when it premiered back in 1949, *On the Town* was considered a landmark motion picture. More than any other film to that date, it removed the musical out of the confines of a studio and set it into the real world.

The Broadway rendition of this engaging entertainment had been derived from an inventive ballet entitled *Fancy Free,* choreographed by Jerome Robbins to the music of Leonard Bernstein. Betty Comden and Adolph Green, who were then performing in musical revues and considered Bernstein their closest friend, were enthralled with the ballet and devised an amusing story built around its basic premise—three sailors on shore leave. They presented the idea to director George Abbott, who commissioned them to write the book and lyrics for this proposed musical comedy venture and also set Bernstein to do the music.

Produced by Oliver Smith and Paul Feigay, with direction by Abbott and choreography by Robbins, *On the Town* opened at the Adelphi Theater in late December of 1944. John Battles, Chris Alexander, and Adolph Green played the three tars, while Nancy Walker, Sono Osato, and Betty Comden were the gals they pursued and won during their day-long excursion in New York City.

From the opening moments of the show when the three sailors romped on stage singing "New York, New York," audience and critics alike were aware that they were watching something special. Bernstein—with his background in serious music—gave the popular musical theater a new, unusual, pulsating sound in *On the Town,* one which was for many years afterwards considered avant-garde. And Jerome Robbins' modernistic choreographic designs for such sequences as "Miss Turnstiles," "Times Square," and "Imaginary Coney Island" were forerunners of what he would conceive later for his unforgettable *West Side Story.* The book itself was a light, campy piece of fluff, but it was perfectly suited to the master plan.

"*On the Town* is the freshest and most engaging musical show to come this way since the golden day of *Oklahoma!*" said Lewis Nichols in the *New York Times.* He then went on to call it "a perfect example of what a well-knit fusion of the respectable arts can provide for the theater . . . . It is an adult musical show and a remarkably good one."

MGM learned of *On the Town* while it was still in its formative stages and, in one of the first pre production deals ever made, bought the screen rights to the yet-to-be-produced play for $250,000. Shortly after the show opened at the Adelphi, studio head Louis B. Mayer journeyed east with two of his top executives and saw the play. He returned to Los Angeles regretting the fact that he'd ever heard of it. As far as he was concerned, *On the Town* was totally unsuitable for movies. The project was shelved.

Near the end of the decade, MGM contractee Gene Kelly let it be known to producer Arthur Freed, who would then be his champion with the MGM brass, that he wanted to direct one of his own films—particularly the dance segments. The property he had in mind was *On the Town*. Stanley Donen would work on the film with him, handling staging of the non-choreographic segments, and Comden and Green would do the screenplay.

Kelly wanted to try something new with the musical numbers in *On the Town*. It was his idea to shoot the film's exteriors away from the studio—right on location in New York City. "It was tough getting them to let me shoot in New York," the actor/dancer told *Entertainment World* in March of 1970. "I had to stamp my foot and act like a movie star. The biggest reason they let me shoot there was that Arthur Freed believed in these silly young people—Comden, Green, and myself—who said that we know if we get off a ship in the Brooklyn Navy Yard and we sing the opening number it would be right. When the executives fought us, we said, 'Nelson Eddy was in a canoe with Jeanette MacDonald and he started singing.' They said, 'That's different. That's something people are used to.' And they were right. They had the same kind of frame of reference as the opera."

For all his "foot-stamping," Kelly wound up with a $1.5 million budget, a forty-six day shooting schedule, and only five days of location filming in New York. He utilized this time, for the most part, to shoot the opening "New York, New York" number in quick—"get the shot when you can"—snatches all over town (Rockefeller Plaza, Wall Street, the Statue of Liberty, and so forth). "The executives thought it was all worthless anyway," said Kelly. "The studio fought because it was expensive. They can control it on the backlot. Today they still would much rather do the shot in the backlot. But by shooting it on the streets and in the Brooklyn Navy Yard, I was trying to gain a whole new look, a whole new feel, and a whole new proscenium for the motion picture musical."

Joining Kelly in the cast of *On the Town* was Frank Sinatra, with whom he'd co-starred in *Anchors Aweigh*, and Jules Munshin and Betty Garrett, who'd been with both of them in *Take Me Out to the Ball Game*. Ann Miller and Vera-Ellen played the other two girls. Vera-Ellen, who was Kelly's dancing partner in the film, had previously worked with him in the "Slaughter on Tenth Avenue" ballet in *Words and Music*.

"It wasn't planned that we girls go on the New York location," recalls Betty Garrett. "After all, we were only needed for one final shot at the Brooklyn Navy Yard—and the studio could have saved money by using stand-ins for us. But Ann Miller went to Arthur Freed and talked him into letting us come along. We had a lot of fun that week."

For those who fondly remembered the stage version of *On the Town*, Metro's rendition must have created some disappointments, since several of Bernstein's best musical moments were deleted. Neither Freed nor the MGM top echelon had ever been over-enthusiastic about the play's score, and besides, if new songs were added, those publishing rights would be owned by the studio. Hence, gone was the beautiful "Lonely Town," and the dazzling ballet, "Gabey in the Playground of the Rich"—both of which had evoked considerable comment from stage critics. Indeed, all that remained from the original score was the opening "New York, New York"; a comic number, sung by Sinatra and Miss

On the Town. **Frank Sinatra, Jules Munshin, and Gene Kelly atop the RCA Building in the opening "New York, New York" sequence.**

On the Town. Vera-Ellen dances the "Miss Turnstiles" number.

On the Town. Jules Munshin and Ann Miller.

On the Town. **Frank Sinatra and Betty Garrett.**

Garrett, "Come Up to My Place"; and the music for the "Miss Turnstiles" and "A Day in New York" ballets. Substituted were six new songs composed by the film's associate producer, Roger Edens, with lyrics by Comden and Green. All the fresh tunes were pleasing to the ear, though none would become standards.

The best of the Edens numbers were a title song, performed by the six principals as they strolled through the New York streets on Metro's backlot; "Prehistoric Man," a wild romp of song and tapdance in which the sailors and the Misses Garrett and Miller cavort about in the Museum of Natural History; and the amusing "You Can Count on Me," staged in a restaurant setting with the aforementioned five players plus zany Alice Pearce.

More than anything else, *On the Town* was a film devoted to the dance—particularly modern and ballet. Kelly's choreography in the "Miss

On the Town. **Gene Kelly.**

On the Town. Gene Kelly (c) with Sinatra and Munshin dancing substitutes in the "A Day in New York" dream sequence.

On the Town. **Vera-Ellen and Gene Kelly.**

Turnstiles" (a showcase for Vera-Ellen), "A Day in New York," in which he was joined by dancers substituting for Sinatra and Munshin, and "New York, New York" gave movie audiences an exciting alternative to the confined heel kicking, tap-style dance numbers they'd been treated to over the years. Certainly, without the success of this pioneer effort, Kelly would never have gotten his shot at *An American in Paris* or *Singin' in the Rain.*

Plotwise, there was not much difference between the stage book and the screenplay, although Betty Garrett claims, "The stage play was much more risque."

Bosley Crowther wrote in the *New York Times:* ". . . the over-all picture flits and frolics with the same carefree delight as did the popular original—and with equal originality, too. Gene Kelly and Stanley Donen, who directed under the eye of Arthur Freed, have actually found some fresh capers for screen musical comedy. They have cleverly liberated action in the manner of the musical stage and they have engineered sizzling momentum by the smart employment of camera techniques."

The movie grossed in excess of four million

dollars, and won the 1949 Oscar for best scoring of a musical film.

Dated as the Arthur Freed production may have become, audiences still pack motion picture revival houses whenever it is shown. They laugh at the corny comic lines and applaud after each and every number. *On the Town* might be a movie milestone whose innovations have been far surpassed by later filmmakers—including Gene Kelly himself—but it still remains good entertainment.

2

Annie Get Your Gun

(1950)

Inconceivable as it might seem, Irving Berlin was not the first composer chosen to write the score for *Annie Get Your Gun*. Producers Rodgers and Hammerstein had Jerome Kern in mind for that project. Then, following Kern's sudden death, they turned to Berlin who, after some procrastination, agreed to take on the assignment. His decision was a wise one. This musical of Americana and its legends turned out to be the most successful Berlin ever wrote for Broadway, and included possibly his most enduring song apart from "White Christmas."

"Annie" was Annie Oakley, the famous lady sharpshooter of the old west. The carefully crafted book by Herbert and Dorothy Fields introduced her during her backwoods days when she was discovered by Buffalo Bill. A dead shot, she joins Colonel Cody's wild west show, falls in love with her handsome marksman rival, Frank Butler, and soon supplants him as the show's star. Butler, his ego bruised, quits and joins a rival extravaganza headed by Pawnee Bill. Later, after the two floundering shows merge, Annie gets her man back when she deliberately loses a shooting match.

Annie Get Your Gun debuted on Broadway on May 16, 1946, at the Imperial Theater. Joshua Logan directed, Helen Tamiris handled the dances, and Jo Mielziner designed the colorful sets and lighting. The cast was headed by the powerhouse Ethel Merman as Annie Oakley, and tall, lanky Ray Middleton as Butler. For both performers, this grand production was perhaps the sterling moment in their careers.

Few musical shows have produced as many engaging and memorable songs as were present in the Irving Berlin score. The big one, of course, was "There's No Business Like Show Business," which has since become the unofficial anthem of the entertainment world. Interestingly, Berlin felt a bit insecure with that number. He wasn't sure that it worked well within the show and suggested that it be dropped. Richard Rodgers, fortunately, disagreed. The song remained.

Aside from "Show Business," the score included a grand array of musical moments. "Doin' What Comes Natur'lly" was Annie's amiable description of the hillbilly life, "Anything You Can Do" was a rollicking boasting/bickering match between Frank and Annie, and in "They Say It's Wonderful" the couple speculated on what love is all about. Frank had some fine solo moments with "The Girl That I Marry," "My Defenses Are Down," and, with a feminine chorus, "I'm a Bad, Bad Man." Annie took center stage when she sang "You Can't Get a Man With a Gun," "I Got Lost in His Arms," "I Got the Sun in the Morning," "Moonshine Lullaby" (with her younger brothers and sisters), and, during her initiation into the Sioux tribe, "Im' an Indian, Too." Two of the show's secondary characters, Tommy Keeler and Winnie Tate (played by Kenny Bowers and Betty Anne Nyman) had a cute number in "Who Do You Love, I Hope?" and Buffalo Bill (William O'Neal) led an ensemble with the opener, "Colonel Buffalo Bill."

"All that is to be wished for in a musical comedy is to be found at the Imperial," said Howard Barnes in the *New York Herald Tribune*. "The new Rodgers-Hammer stein production has every hallmark of distinction. The Herbert and Dorothy Fields libretto is both lively and funny. The Irving Berlin songs form a fascinating web of wit and melody for the action." And of the play's star, John Chapman in the *Daily News* wrote, "Miss Merman can wrap 'em [songs] up and put 'em away with the zing, the punch and the instinct for showmanship she first revealed when she came out upon the Alvin stage and hurled 'I Got Rhythm' [in the 1930 Gershwin show, *Girl Crazy*] at all us delighted discoverers."

Annie Get Your Gun played 1,147 performances in its initial run. A national company of the musical headlined Mary Martin, who later starred on Broadway for Rodgers and Hammerstein in *South Pacific* (1946).

Metro-Goldwyn-Mayer purchased the film rights to *Annie Get Your Gun* in 1947 for a reported price of $650,000. Arthur Freed, master producer of the movie musical, was to guide the film, which would star Judy Garland. Sidney Sheldon wrote the screenplay, opening the scenes up considerably to take advantage of the different medium, but other wise remaining close to the original Fields book. Eight Berlin tunes, however, were dropped from the score—presumably to keep the picture's running time under two hours. Among those to go were "Moonshine Lullaby," "I Got Lost in His Arms," "Who Do You Love, I Hope?" and "Colonel Buffalo Bill." Another tune, "Doin' What Comes Natur'lly," had some lyric changes due to censor ship, and Berlin also wrote a new song for the movie—a pretty ballad called "Let's Go West Again"—which was ultimately dropped.

To support Miss Garland, Metro cast several of its top character players in various roles. Frank Morgan was Buffalo Bill, Edward Arnold played Pawnee Bill, J. Carrol Naish was Sitting Bull, and Keenan Wynn was Charlie Davenport, the show manager. Howard Keel, a virtual newcomer to films who'd been starring in the London company of *Oklahoma!*, was signed by the studio to co-star as Frank Butler. The actor was a fantastic baritone and the studio figured that, if he caught on with the public, they had a perfect leading man for any number of their future musical entries.

"I spent my first eight months at MGM doing nothing," says Keel, "and they were paying me $850 per week. Louis B. Mayer insisted that I make my Hollywood debut in a musical, so, since Garland was busy on another picture, I just sat-around and collected my pay."

With Busby Berkeley set to direct and Robert Alton assigned as choreographer, Garland and Keel began pre-recording the show's musical numbers in late March of 1949. But something was amiss. The people around her were aware that Judy's singing was not up to par. Annie Oakley was a character totally foreign to anything she'd ever played before. She was unsure of herself and, according to one insider, "was in awe of Merman." Missing from her performance was that elusive bit of magic she instilled into all her songs.

"You could see that she wasn't right," reflects Keel. "As the picture progressed, she got worse. I suppose she was on uppers and downers just to keep going."

The problems on *Annie* never seemed to end. On the third day of shooting, Keel fell off his horse and broke his ankle. He wore a walking cast for several weeks while the company shot around him.

A few days after that accident, Judy started filming her scenes. Observers recall her extreme nervousness and how she was unable to complete one of the fairly elaborate production numbers, "I'm an Indian, Too," in less than six days. She began leaving the set early, telling Berkeley that she was ill. The same excuse was used when she decided not to come in on particular days, or when she arrived on the set hours late.

Berkeley was fired as director at the end of twelve days of filming. Freed had felt that his concept of the film was totally off base, and as a replacement the producer set Charles Walters.

Had Walters been on the movie from the start, perhaps Judy Garland might have fared better, since

Annie Get Your Gun. Louis Calhern, Betty Hutton, and Howard Keel.

Berkeley was not her favorite director and the two seemed to clash. But realistically it's doubtful that anyone could have truly helped her since the lady was simply too overworked and exhausted to continue. Following a confrontation with the studio head office, Garland was put on suspension. Her continued absences from the set were costing too much. Shortly thereafter she entered a Boston hospital for a complete rest.

Four weeks and nearly a half million dollars had been spent on *Annie.* No usable footage was in the can. Metro had to recoup its investment, so the brass began seeking a replacement for the star. Betty Garrett, whose contract with the studio had recently expired, was approached: "I didn't want to sign another seven year contract. Besides, they were offering less money than they'd paid me before."

Arthur Freed looked elsewhere, and after considerable negotiation a deal was made to borrow Betty Hutton from Paramount for $150,000. The producer would have to wait until Miss Hutton finished work on her current film, in which she was co-starring with Fred Astaire, but that didn't pre-

Annie Get Your Gun. J. Carrol Naish, Betty Hutton, and Louis Calhern.

Annie Get Your Gun. Betty Hutton (r) and player.

Annie Get Your Gun. Howard Keel, Louis Calhern, and Betty Hutton.

Annie Get Your Gun. Keenan Wynn, Howard Keel, and players.

sent him with too great a problem since his production had shut down and his cast was under contract to MGM anyway.

Two more changes in personnel occurred before principal photography began again on October 10, 1949. George Sidney took over as director from Charles Walters, and decided—considering the talents of his new leading lady—to emphasize comedy over dance numbers. Following the death of Frank Morgan, Louis Calhern assumed the role of Buffalo Bill.

Annie Get Your Gun cost MGM over $3.75 million dollars, but the finished product, which debuted in early 1950, was well worth the time, money, and expense involved. In many respects, the play worked better on film than it had on the stage. Sidney captured the glamour and excitement of the wild west show, and, with the screen's greater scope, turned his musical numbers into overwhelming bodies of activity and color, beautifully tinted and lustily staged. The finale featuring the combined Buffalo Bill/Pawnee Bill shows riding and performing to "There's No Business Like Show Business" was magnificent spectacle as only the movies can provide.

Though certainly not the equal of Miss Merman, Miss Hutton was excellent as the slightly uncouth Annie and, thanks to her director's strong hand, was for the most part able to subdue the flagrant screen mugging for which she'd achieved a dubious reputation. It was Keel, however, who benefited most from the show. Women swooned over him. Critics tagged him "a Gable who sings." A new star was born.

Said *New York Times* critic Bosley Crowther: "Into this screen translation of the Berlin-Fields musical comedy hit, . . . the West Coast producers have loaded more people, more gaudy Wild West show, more horses, more guns and more clay pigeons than could ever be assembled on a stage. With diligence and fidelity—and with Technicolor, of course—they have splashed on a limitless canvas the tale of Annie Oakley, the crack shot. They have used a good cast of actors, they have seen that the music is done well and they've got a new chap to play Frank Butler, Annie's boy friend and rival, who is a pip."

Including its 1956 re-release, the picture has grossed in excess of eight million dollars to date, and is considered by many historians to be one of the most satisfying musicals to emerge from the Culver City studio during the decade. With that fabulous Berlin score, how could it not be?

3
Gentlemen Prefer Blondes
(1953)

On December 8, 1949, Carol Channing stepped onto the stage of New York's Ziegfeld Theater and proclaimed that "Diamonds Are a Girl's Best Friend." The next morning, this vivacious wide eyed platinum blonde with the shrill baby voice was Broadway's bright new musical comedy star.

The show was *Gentlemen Prefer Blondes,* with book by Anita Loos and Joseph Fields, based on Miss Loos's novel and subsequent stage comedy. Leo Robin did the lyrics, Jule Styne composed the music, John C. Wilson directed, and choreography was by Agnes deMille.

Miss Loos's story was a wild, witty, satirical homage to the jazz age—the 1920s. As flapper Lorelei Lee, Miss Channing personified the era and the gals who played their ways through it.

The musical opens on an ocean liner. Showgirl Lorelei and her pal Dorothy Shaw (Yvonne Adair) are going to France to escape Prohibition. Lorelei has a wealthy fiancé, Gus Edmond, who sees her off.

In France, Dorothy falls for a rich American, Henry Spofford (Erich Brotherson), while Lorelei becomes involved with Sir Francis Beekman. Her main interest in Beekman is to borrow the money to buy his wife's diamond tiara. The play's major complications arise when Lady Beekman learns of her husband's flirtation with Lorelei, then starts legal action to recover the jewels. Gus, in the meantime, discovers Lorelei with still another millionaire and breaks their engagement.

As with almost all fluff y musicals of this ilk, *Gentlemen Prefer Blondes* ends happily for everyone involved. Lorelei's legal problems get solved, and she gets her Gus.

"Diamonds Are a Girl's Best Friend," Lorelei's amusing statement of her life philosophy, was of course the musical's show-stopper. But, there were also two other numbers that achieved a mild popularity—Lorelei's autobiographical ballad, "A Little Girl from Little Rock," and "Bye, Bye, Baby."

Whatever the production's other attributes, *Gentlemen Prefer Blondes* belonged to Miss Channing, as Brooks Atkinson so stated in the *New York Times:* "She has something original and grotesque to contribute to every number. She can also speak the cock-eyed dialogue with droll inflections. Her Lorelei is a mixture of cynicism and stupidity that will keep New York in good spirits all winter ." And, in the New York *Daily News,* John Chapman described the play as "a hilarious libretto. . . . Everything about *Gentlemen Prefer Blondes* is precisely right."

The production played 740 performances in its initial engagement.

Paramount Pictures had produced a 1928 silent

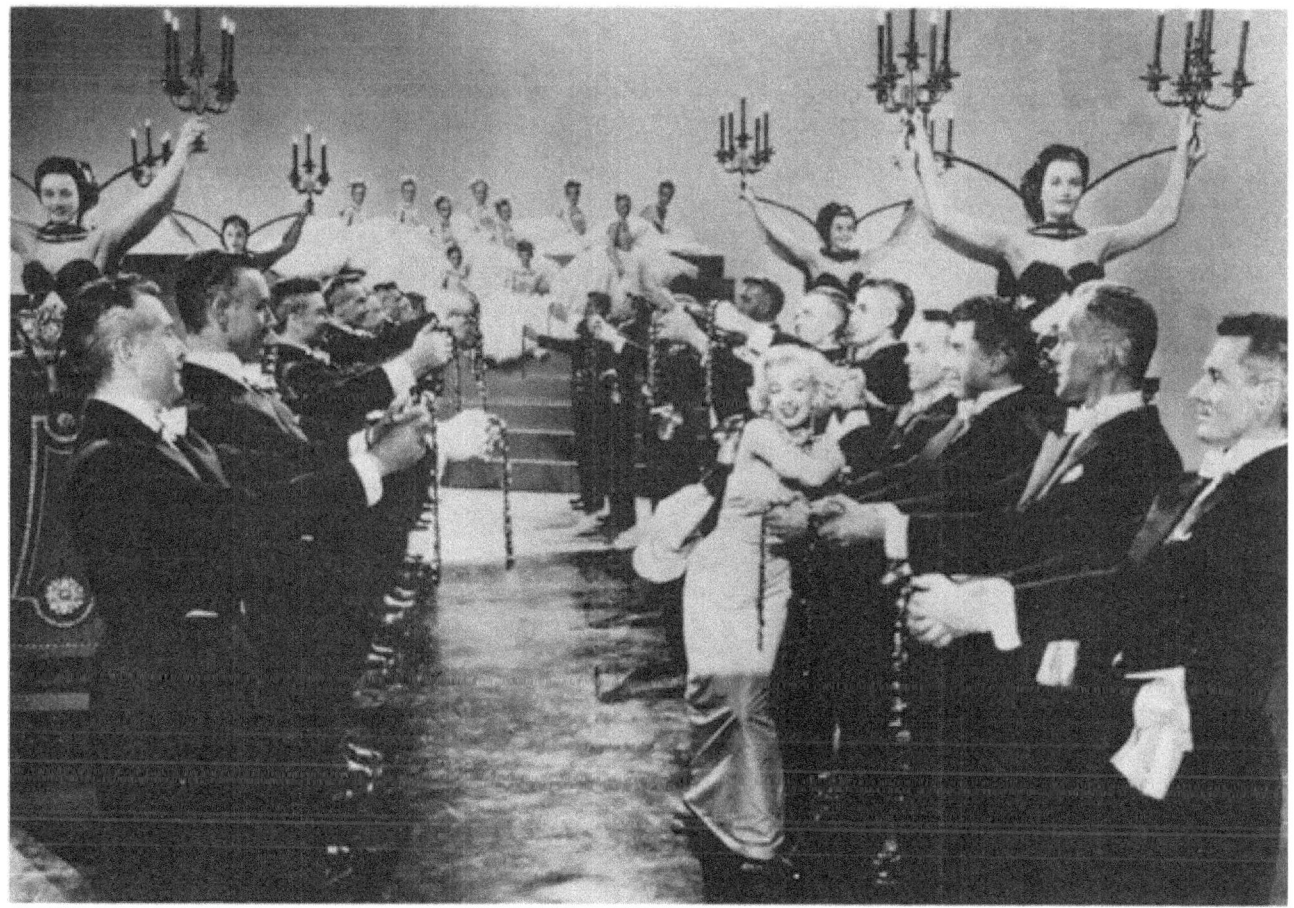

Gentlemen Prefer Blondes. Marilyn Monroe sings "Diamonds Are a Girl's Best Friend."

version of Anita Loos's original play and, as a result, maintained a 16⅔ % interest in the musical's film rights. Herman Levin, who co-produced the 1949 hit with Oliver Smith, bought out the studio's interest in 1951 and later that year sold the rights to Twentieth Century-Fox for $150,000.

Assigned to produce the project for Fox was Sol C. Siegel, who'd previously handled the movie adaptation of Irving Berlin's *Call Me Madam.* "I'd seen the show in New York," Siegel says, "and, although I found it entertaining, there was not much substance to the plot. It was too thin to work in a film. Broadway musicals, back then particularly, were notorious for their weak story lines.

"It was not a property I would have chosen for myself, but Zanuck asked me to do it as a favor."

With Charles Lederer doing the screenplay, Siegel made some drastic changes in the original Fields/Loos script. First the story was updated. Instead of the twenties, *Gentlemen Prefer Blondes* now took place in the present—1953. Second, the role of Dorothy, Lorelei's man-crazy girl friend, was enlarged so that both parts were relatively equal in size. The basic plot elements remained stable, except that some new comedic situations were devised to go along with the aforementioned alterations. For example, instead of falling for a millionaire, Dorothy now becomes smitten with a private detective who is trailing Lorelei on behalf of the blonde 's rich boy friend. The wealthy Mr. Spofford is no longer an adult, but a precocious young lad, traveling alone on the liner except for his valet. One of the picture's more amusing scenes, in fact, occurs when an astonished Lorelei, who'd set her materialistic sights on Spofford, discovers his true age.

Once Fox had announced its purchase of *Blondes,* speculation began as to who would play Lorelei. The "smart" bettors put their money on Carol Channing to recreate her part for the screen, al though others claimed that studio contract star Betty Grable would get the choice assignment. "I never considered Grable," reports Siegel, "but there were conversations with Miss Channing. We wanted to test her, but her agent refused to let her test unless we gave her a Fox contract beforehand. On that basis, I refused and went with Marilyn Monroe, whom we already had under contract."

The studio borrowed Jane Russell from RKO to co-star as Dorothy in the 1953 Technicolor production, and, although Lorelei was certainly the pivotal character in both the stage and screen versions of the Loos work, the brunette actress was awarded top billing. The pairing of the two visual delights, however, was exciting casting. The girls worked well together.

Also in the company were Charles Coburn as Beekman, Tommy Noonan as Lorelei's fiancé, Elliott Reid as the private eye, and little George "Foghorn" Winslow as Spofford. The director was Howard Hawks.

"Jack Cole did the choreography for the picture, " says the producer, "but it was his assistant, Gwen Verdon, who walked Marilyn through the 'Diamonds Are a Girl's Best Friend' number."

"Diamonds," "Bye, Bye, Baby," and "A Little Girl from Little Rock"—retitled *"Two* Little Girls from Lit tle Rock" to accommodate Miss Russell—were the only Styne/Robin numbers retained from the stage. Two new tunes-"When Love Goes Wrong" and "Anyone Here for Love"—were added. They were written by Hoagy Carmichael and Harold Adamson rather than the original composers. Siegel: "Hawks had worked with Carmichael before and specifically asked that he do the new songs."

Gentlemen Prefer Blondes was an okay musical film that, through Hawks' sure direction, preserved the spirit, if not the text, of the stage production, providing audiences with some cute moments. The five musical numbers were nicely executed by the well-endowed and beautifully attired (courtesy of Travilla) Misses Monroe and Russell. Two or three more songs would have helped the picture, since Charles Lederer's script—despite the efforts of Siegel—was still pretty thin and could have used the diversion.

Newsweek: "The film may be easily viewed, and easily is the word, either as a form of drama or as a piece of erotic anthropology . . . . Miss Monroe's 'Diamonds' number is a superb example of the way

Gentlemen Prefer Blondes. Jane Russell and Marilyn Monroe sing "When Love Goes Wrong."

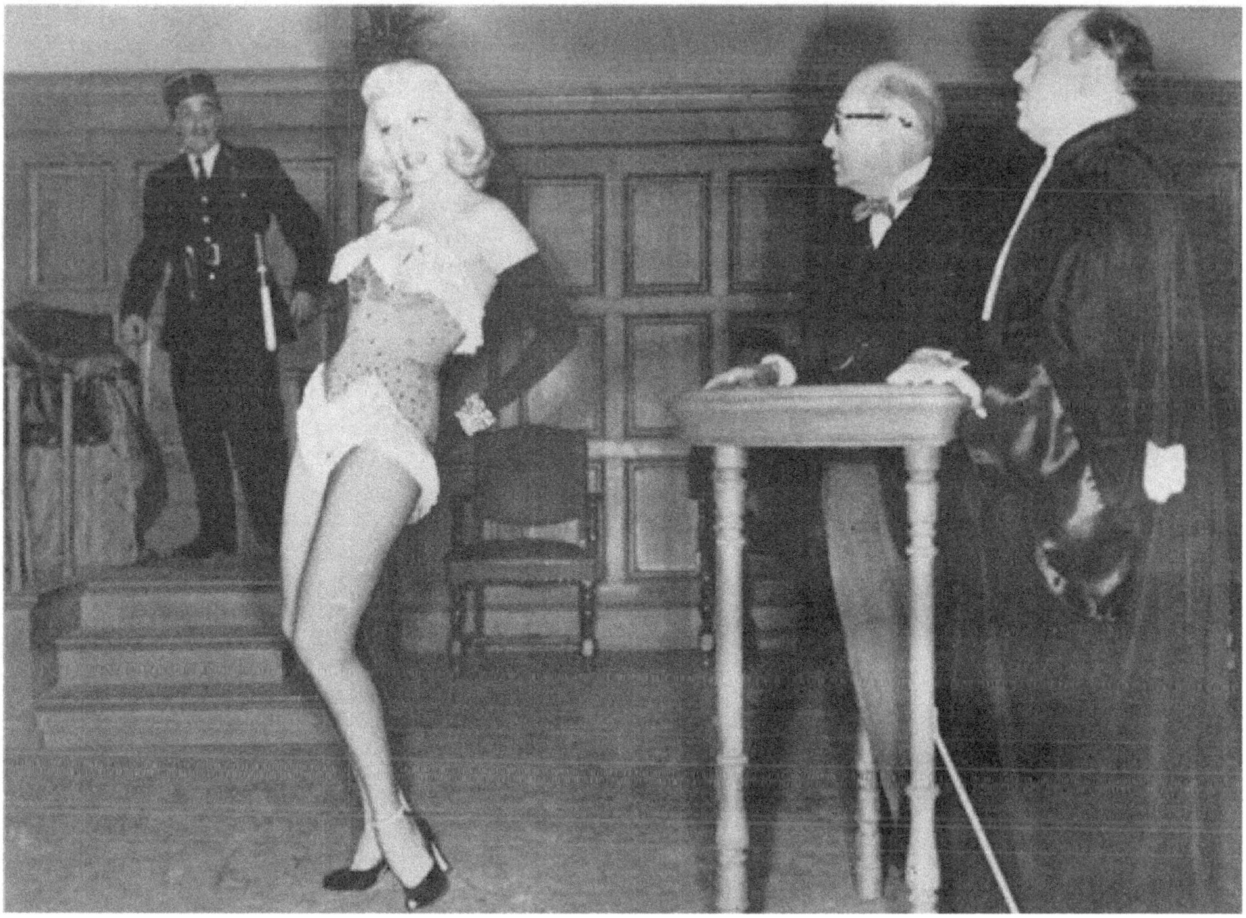

Gentlemen Prefer Blondes. Jane Russell shocks the courtroom.

Gentlemen Prefer Blondes. Elliott Reid and Charles Coburn.

the camera itself can become one of the most expressive of the dancers."

Blondes, which earned domestic film rentals of $5.1 million, gave a major career boost to Marilyn Monroe, who up to that time had not been cast too astutely by her employers at Fox. It was in this picture that her sultry, almost too innocent screen Personality—which would become the basis of numerous parodies—was firmly established. Indeed, if it were not for her presence and her sensual rendition of "Diamonds Are a Girl's Best Friend," it's doubtful if the movie would be as well remembered as it is today.

4
Kiss Me, Kate
(1953)

Cole Porter had given several fine shows *(The Gay Divorce, Anything Goes, Leave It to Me)* to the musical theater, but as the 1940s drew to a close, Broadway insiders considered him virtually washed up. A horseback riding accident back in 1937 had left the composer's legs crippled and, possibly as a result of his continual pain and suffering, he hadn't written a successful show or tune for years. That downward trend, however, was about to reverse itself.

The production that catapulted Porter back into the upper echelon of "hot" Broadway composers was *Kiss Me, Kate,* a show that is often mentioned in the same breath as *Oklahoma! and Annie Get Your Gun.* Produced by Saint Subber and Lemuel Ayers, it had a lively book by Bella and Samuel Spewack that was inspired by Shakespeare's bawdy comedy, *The Taming of the Shrew.* John C. Wilson directed and Hanya Holm choreographed the show, which opened on December 30, 1948 at the New Century Theater.

Kate is a play within a play. It tells of a theatrical company in Baltimore that is presenting a musical production of the Shakespeare comedy. Stars of that troupe are a divorced couple, Fred and Lilli Graham (Alfred Drake and Patricia Morison), who, despite their differences, still love each other. But Fred has been pursuing the show's ingenue, Lois Lane (Lisa Kirk), a flirt and the girl friend of Bill Calhoun (Harold Lang). Bill is a gambler who owes a bundle to some local gangsters. Two hoods arrive at the theater just prior to performance to collect the debt.

The continuity of the Spewack book is loose, switching back and forth from the jealousies and frantic misunderstandings backstage to the onstage action in old Padua. Not so coincidentally, both stories have their parallels. At the end of the musical comedy, Fred and Lilli are back together, Lois has her Bill, and the two comic gangsters are seeking new employment—their boss having met with a sudden "accident."

Always the master of the clever, slightly naughty lyric, Porter outdid himself with *Kiss Me, Kate,* delivering songs in a variety of musical styles. "Wunderbar" was a spoof of the old-fashioned European operetta, Lois gave Bill a rational excuse for her flirtations with other men in "Always True to You in My Fashion," "Too Darn Hot" was a sharp commentary on Baltimore's weather, and "So in Love" was Fred and Lilli's romantic ballad. The thugs also had a great musical moment and, in

Kiss Me, Kate. **Kathryn Grayson, Tommy Rall, Ann Miller, and Howard Keel sing "We Open in Venice."**

fact, stopped the show with their amusing exit, "Brush Up Your Shakespeare." Composer Porter let his well-crafted material spread into the *Shrew* portion of the play as well. "We Open in Venice," "I Hate Men," "Were Thine That Special Face," and "Where Is the Life That Late I Led" were unique gems, written to fit the Shakespearean manner and idiom.

Richard Watts, Jr., of the *New York Post:* "To Mr. Porter, the new musical comedy must be a particularly gratifying success, since there had been dark rumors abroad that the eminent composer had lost some of his old-time power. There is no sign of any such decline in either the music or the lyrics of *Kiss Me, Kate*. There are some seventeen numbers, and all of them are worth listening to. There is no one song that struck me as standing out above the rest on first hearing and the lyrics are probably better than the tunes, but it all adds up to a vastly engaging score."

The actors, choreography, direction, and entire production concept were roundly praised by reviewers. *Kiss Me, Kate* played 1,077 performances and garnered the Tony award for the best musical of the season.

MGM bought the film rights to Porter's show and assigned Jack Cummings to produce, Dorothy Kingsley to script, and George Sidney to direct. The production was to be one of the studio's important musical projects for 1953, and, just to be sure that all bases were covered, it was to be shot twice—once in a regular format, and again in the faddish 3-D process. "My cameraman and I had to compose every shot three different ways [for three different aspect ratios] at the same time," Sidney told the *Los Angeles Times.* "What would be good for one width would not be good for another. It was tricky, but we got around it by building more tops on sets, more floor and more sets in forced perspective to enhance

Kiss Me, Kate. Howard Keel and Kathryn Grayson.

Kiss Me, Kate. Tommy Rall, Ann Miller, and player.

Kiss Me, Kate. Howard Keel sings "Where Is the Life That Late I Led?"

the depth. The wider the screen, you see, the narrower; we had to compensate for those cut-off tops and bottoms . . . . Even if you see *Kate* flat, you'll notice that it seems to have more depth than the ordinary movie." When *Kate* was finally re leased, 3-D was on the way out and the movie did play most engagements flat.

With one major exception, producer Cummings set about to cast his prestige project with some of Metro's top stars. Kathryn Grayson played Lilli, Ann Miller, who'd been under consideration for the assignment on Broadway, was Lois, Tommy Rall played the part of Bill, and Keenan Wynn and James Whitmore were the two thugs. Yet for the all important male lead, Cummings was intrigued with the idea of using an outsider—Laurence Olivier. The fact that the great British actor's singing voice would have to be dubbed was compensated by the knowledge that his presence would certainly add a considerable elegance to the picture.

Howard Keel, the studio's baritone star of such musicals as *Annie Get Your Gun* and *Show Boat*, wanted the part of Fred Graham. His last few films at Metro had been lemons, and he felt that this major role would put new life into his career. With director Sidney and Miss Grayson backing him, Keel began his assault on Cummings's firm stance that he was wrong for the role. He won himself a test and, on the

Kiss Me, Kate. Ann Miller dances the "Tom, Dick or Harry" number with Bobby Van, Tommy Rall, and Bob Fosse.

Kiss Me, Kate. Howard Keel and Kathryn Grayson.

basis of his performance in that short piece of film, was awarded the assignment.

Although Miss Kingsley's screen adaptation was essentially faithful to its parent material, there were some interesting changes from the original. On film, the entire show takes place in New York rather than in Baltimore, and in an opening sequence which the *New York Times* called "one of the silliest and clumsiest beginnings that ever a musical has had," Cole Porter himself (played by Ron Randell) tries to talk Fred and Lilli into doing his play.

Music-wise, several songs from the original show had their lyrics changed due to censorship problems, others were dropped, still another number—"From This Moment On"—was snatched from Porter's *Out of This World* and stuck into the movie score, and "Too Darn Hot," which opened the second act and was sung by a relatively minor male character, was moved to the first scene and given to Miss Miller to build up her part. She endowed it with a jazzy interpretation.

The opinion of many fans of the movie musical was that Hermes Pan had choreographed perhaps the most dazzling, inventive dance numbers seen outside a Kelly or Astaire film. One spectacular design featured Miss Miller, Rall, and dancers Bobby Van, Bob Fosse, Carol Haney, and Jeanne Coyne leaping about on the arch-enclosed Shakespearean set to the notes of "From This Moment On."

As with the play, the entire production—costumes, settings, direction, and acting—was applauded, with Howard Keel perhaps coming off with the lion's share of the good performance notices. According to *Hollywood Reporter,* "He sings wonderfully with a voice that will thrill any audience. He looks good, acts the part without a flaw, and comes up with a performance to far exceed anything he has done heretofore . . . ."

And of the movie itself the trade paper said, "In this picture, they [MGM] have blended wonderful, colorful sets into optical orgies, using them as backgrounds for gay musical numbers enacted by artists who have no comparable values out of this studio."

Despite the press approval, *Kiss Me, Kate* was not a boxoffice hit, even though it is today cherished by fans at movie nostalgia revivals. Maybe, back in 1953 when it was released, the idea of Shakespeare in any form was just too highbrow to attract a mass audience, particularly when television was still a novelty and, of course, free.

5
Oklahoma!
(1955)

Almost everyone connected with the original stage production of *Oklahoma!* expected that show to be a box office flop. Almost everyone except composer Richard Rodgers, that is.

The doubters, who included Rodgers' collaborator, Oscar Hammerstein II, argued a convincing case. The show had no stars, it was based on a play that had failed in 1931, and director Rouben Mamoulian had worked only once before in musical theatre. He'd staged an opera in 1935 that was a financial *bomb—Porgy and Bess.* If these factors weren't enough, *Oklahoma!* was violating most of the concepts and established rules of musical comedy.

Lawrence Langner and Theresa Helburn of the financially troubled Theatre Guild had conceived the idea of turning Lynn Riggs's folk play, *Green Grow the Lilacs,* into a musical. They approached Rodgers and Hart to do the score, but Lorenz Hart was ill and not interested. Thus lyricist/adapter Hammerstein was ultimately brought into the project. The immortal partnership was born .

The new collaborators decided to defy convention with their folk play and experiment with some original ideas that had never before been tried in musical theater. Instead of opening on an ensemble of beautiful chorus girls, the play begins with Aunt Eller (Betty Garde) on stage alone, churning butter, while the hero, Curly (Alfred Drake), strolls in a few moments later singing solo. Agnes deMille, a leading figure of the ballet, staged the dance numbers as American ballets, rather than in the staid style of heel-kicking choreography normally expected in this type of light entertainment. Of greatest importance was the fact that most of the overdone cliches germane to musical comedy were thrown out in favor of letting the script determine what method would be used. Little was sacred. In *Oklahoma!* all songs, dances, and comedy bits were dictated by character and plot, not by a producer's whim that such elements were required at a certain point to fit a predetermined format. The result was a zesty musical that—while featuring a myraid of fresh, poetic songs that generated a feel of the outdoors and have since become classic ("Oh, What a Beautiful Mornin'," "The Surrey with the Fringe on Top," "I Cain't Say No," "People Will Say We're in Love," "Pore Jud," and, of course, the title tune)—finally gave audiences a musical production that was a cohesive dramatic entity. Its appearance on the scene completely revolutionized the art form.

Set near the turn of the century, *Oklahoma!* has a

Oklahoma!. Shirley Jones and Gordon MacRae.

Oklahoma!. Shirley Jones.

very simple plot. Curly, a cowboy, and Eller's niece, Laurey (Joan Roberts), love each other. Jud (Howard da Silva), a sinister hired hand on the ladies' farm, is also taken with Laurey and unsuccessfully courts her. At the couple's wedding, he challenges Curly and is accidentally killed in a fight. Curly is exonerated and the couple go off on their honeymoon.

A number of amusing and well-drawn folksy characters gave the play its true warmth and texture. Naive cowboy Will Parker (Lee Dixon), who sings "Kansas City," is smitten with the perplexing Ado Annie (Celeste Holm). She in turn is interested in Persian peddler Ali Hakim (Joseph Buloff), a self styled Casanova. Then there is Annie's father, Judge Andrew Carnes. He doesn't like Will, and he makes the marriage-shy peddler nervous with his ever present shotgun.

Oklahoma! opened at the St. James Theater in March of 1943. In a typical critical rave, John Anderson of the *Journal-American* described it as "beautiful . . . delightful . . . fresh . . . imaginative." The initial engagement went on to play 2,212 performances, grossing seven million dollars.

It took twelve years for *Oklahoma!* to reach the movie screen. National and world tours, plus revivals, kept the property active during that wait. When the legendary musical did become a film in 1955, it again made history. Produced under the personal supervision of Rodgers and Hammerstein for distribution by the Magna Theater Corporation, the picture had the distinction of being the first shot in TODD-AO, Mike Todd's new wide-screen process that offered audiences the all-encompassing sweep of Cinerama, but with only one projector instead of three. Arthur Hornblow, Jr., was the line producer on the six million dollar project, which was directed by Fred Zinnemann, whose credits included *High Noon* and *From Here to Eternity.* Sonya Levien and William Ludwig did the screenplay.

"I didn't think the movie was as good as it could have been," says Ludwig. "We weren't allowed the opportunity to break it out of the proscenium and take full advantage of the camera.

"Rodgers and Hammerstein had complete control down to the last comma. You really couldn't argue with their success or their control. They

Oklahoma!. Shirley Jones, Gordon MacRae, and Charlotte Greenwood.

Oklahoma!. Gloria Grahame and Shirley Jones.

Oklahoma!. Rod Steiger and Shirley Jones.

Oklahoma!. Gordon MacRae and Shirley Jones.

wanted the film to be as close to their smash stage success as possible.

"Every time Fred Zinnemann tried to open up a scene and use the camera, Hammerstein was there to pull him back. For example, Agnes deMille choreographed 'The Farmer and the Cowman' number *exactly* the way she'd done it on the stage twelve years before. When Fred argued that he wanted to utilize a large camera boom and play the number over the entire set, which filled two sound stages at Metro, he was overruled.

"The only number he was allowed to expand was 'Kansas City,' which we staged outside at the railroad depot and on top of a train."

Ludwig found that he and Miss Levien had trouble in both enhancing and editing Hammerstein's original script: "We tried to expand the character of Jud—flesh him out a bit with short scenes not in the original play—but these were eliminated before filming began.

"Another problem we had was with the play's repetitive lines. On stage it might be necessary to repeat a fact two or three times in order to set up a gag. But, in movies, a close-up can make your point the first time. We wanted to cut some of the repetition, however Hammerstein wouldn't hear of it. He had a tremendous loose-leaf notebook with him that appeared to be a record of 'every' audience reaction at 'every' performance of *Oklahoma!* If I suggested cutting such and such a line, he'd look in his book and say, 'Impossible! That line got a sixteen second laugh in Boston.' He wasn't about to sacrifice anything."

Two inconsequential songs were, in fact, cut in the film version: "Lonely Room," sung by Jud, and the Ali Hakim number, "It's a Scandal! It's an Outrage!"

Virtually every musical performer in Hollywood tested for the roles in *Oklahoma!* "I vaguely remember being impressed with the competition on all levels," says lanky dancer Gene Nelson, who played Will Parker. Gordon MacRae beat out Howard Keel for Curly, newcomer Shirley Jones was Laurey, Charlotte Greenwood played Aunt Eller, Gloria Grahame was Ado Annie, Eddie Albert played Ali Hakim, and Rod Steiger was Jud. (Howard da Silva, who played the part in New York, had been discussed to repeat the role, but this, unfortunately, was the era of the blacklist.) Other cast members included James Whitmore, Barbara Lawrence, Jay C. Flippen, and, dancing for MacRae and Jones in the emotional dream ballet sequence, James Mitchell and Bambi Linn.

The company rehearsed on mock-up sets at the studio for seven weeks, then moved to the San Rafael Valley—thirty-six miles northeast of Nogales, Arizona—to film exteriors. Actual Oklahoma locales had been scouted, but were vetoed after it was decided that there would be little opportunity to capture scenic beauty without running into commercial development.

Oklahoma! began with the audience-embracing TODD-AO camera moving through stalks of golden corn, then picked up Curly riding along the prairie. The eye-catching scenery as he sang "Oh, What a Beautiful Mornin'," plus the exterior "Kansas City" number, and a wild ride on a runaway buckboard with Laurey and Jud were among the few instances in which the movie broke out of its stage origins.

Yet this made no real difference. The play was still the important thing in this near-religious adaptation—thanks to Rodgers and Hammerstein's intense monitoring of the production. They wanted movie audiences to experience the same thrill that had gripped their theater counterparts. Music, dances, comedy bits were almost all intact and just as exciting, colorful, and fun as when they were first presented back in 1943. Agnes deMille's choreography received particular praise. Disagreeing with any detractors, the *Hollywood Reporter* commented, "Far from being content to merely repeat her success in the Theater Guild offering, Miss deMille has created many stage pictures and bits of business especially for the camera. Here the dancing is not an extraneous element of entertainment, but an essential part of the main narrative line."

Although Steiger might have been a little too internal as Jud, and Eddie Albert too American as the Persian peddler, the rest of the casting was superb. MacRae was perfectly cast, cocky without once appearing conceited, and his rendition of the songs was magnificent. Gene Nelson was truly engaging as hayseed Will Parker in what would be his last film musical. The movie propelled winsome, dewy Shirley Jones to major stardom.

Bosley Crowther, writing in the *New York Times,* said, "a full-bodied *Oklahoma!* has been brought forth in this film to match in vitality, eloquence and melody any musical this reviewer has ever seen."

Oklahoma! earned a domestic gross of $7.1

million, and won the Academy Award for best scoring of a musical picture, as well as best sound recording.

William Ludwig: "You never argue with an audience. They're always right. And they loved the movie of Oklahoma!

"But, I can't help wondering—if we had been given the chance to really use the camera—would the picture have been just a *little* better? Would it have had an extra beat? An extra charge of electricity?"

Ludwig recalls that Fred Zinnemann approached him at the premiere and mused, "I wonder if it was worth a year of our lives."

The filmmakers' frustration at being unable to make full use of their medium is certainly understandable. Perhaps they can console themselves with the thought that, despite their limitations, they still created a production that magnified and strengthened all of the charm Oklahoma! had on the stage—a celluloid monument to one of the great milestones in the American theater.

6
Guys and Dolls
(1955)

Damon Runyon was a storyteller who wrote of the mythical land of Broadway and all the horse players, crapshooters, fight managers, and other angle merchants who populated it. He disguised his real-life street characters with such monikers as Harry the Horse, Society Max, Angie the Ox, Nicely-Nicely Johnson, and Nathan Detroit, giving them a highly imaginative language—sans contractions—all their own. "They're a good, kind bunch; lovable types almost to a man," Joseph L. Manckiewicz, who wrote and directed the movie version of *Guys and Dolls,* told *Cue* in 1955. "Even the bad guys are never really evil. Just like the big bad giant in 'Jack and the Beanstalk' who never actually ate anybody, Runyon's mobsters wouldn't hurt a fly."

Guys and Dolls was based on a story by the late columnist, entitled "The Idyll of Miss Sarah Brown." With a score by Frank Loesser, and book by Jo Swerling and Abe Burrows, it opened at the 46th Street Theater in November of 1950. Direction was by George S. Kaufman and choreography by Michael Kidd. It was produced by Cy Feuer and Ernest H. Martin.

The bare story concerns a bet between gambler Sky Masterson (Robert Alda) and Nathan Detroit (Sam Levene), proprietor of the town's "oldest established permanent floating crapgame." Needing a grand to secure a locale for his illegal game, Detroit bets Sky that the latter cannot take Sarah Brown (Isabel Bigley) of the Salvation Army to Havana with him for an evening. A subplot involves Detroit's long-standing engagement to Miss Adelaide (Vivian Blaine), who is becoming impatient to take that walk to the altar.

Masterson promises Sarah that if she will go to Cuba he will guarantee that at least a dozen genuine sinners will attend a prayer meeting at her failing mission. She agrees and, despite herself, becomes enamoured with the flashy man-about-town. Later, when she learns of the bet, Sarah swears never to see Sky again.

The gambler keeps his word to Miss Brown. In a crapgame held in a sewer, he bets his fellow players cash against their attendance at the prayer meeting. He wins, and that night the mission has a full house. At the play's conclusion, Sky and Miss Sarah marry, as do Nathan and Adelaide.

Guys and Dolls was a rich, vibrant musical portrait of Runyon's Broadway. No scene . . . no song . . . no reprise was included in the show unless it enhanced the whole. Well-thought-out numbers

Guys and Dolls. Frank Sinatra, Johnny Silver, and Stubby Kaye in "The Oldest Established Permanent Floating Crapgame" number.

like "Fugue for Tinhorns," in which, amid Broadway's bustling action, three gamblers pick their horses for the day's races, and "My Time of Day," Sky's explanation of his lifestyle, were designed not only to present very listenable songs, but also to develop character, set a mood, or help further the plot in some fashion. Michael Kidd choreographed his faultless dancers, to fast-paced "Runyonland Music," to create an instant portrait of the street people—sidewalk salesmen, touts, pickpockets, hustlers, and constantly frustrated cops.

Several songs, as they say, stopped the show. "The Oldest Established" and "Guys and Dolls," sung by the likes of gamblers Nicely-Nicely Johnson (Stubby Kaye) and Benny Southstreet (Johnny Silver), as well as Nicely-Nicely's confession at the prayer meeting, "Sit Down, You're Rockin' the Boat," are unforgettable. Vivian Blaine scored strongly with tunes such as "Adelaide's Lament," "A Bushel and a Peck," "Take Back Your Mink," and, with Sam Levene, "Sue Me." Robert Alda and Isabel Bigley shared romantic pieces like "I'll Know" and "I've Never Been in Love Before," while Miss Bigley delivered the whimsical "If I Were a Bell" solo. Finally, there was the most popular number to emerge from the Loesser score, "Luck Be a Lady," which Alda sung as he shot dice in the sewer.

William Hawkins, writing in the *New York World-Telegram,* said of the entertainment; "Love for the new musical spread faster last night than fire through dry grass in a high wind . . . *Guys and Dolls* is a show that deserves gold stars in many departments. To my mind the most startling contributions are the affectionately witty idea and musical vernacular of Frank Loesser's score, and Michael Kidd's sharply staged dances."

Guys and Dolls wound up playing twelve

hundred performances before it closed. It won the Antoinette Perry, Donaldson, and New York Drama Critics Circle Awards as the season's best musical.

Hollywood began a furious bidding session for screen rights to Loesser's hit musical. Paramount wanted it for Bing Crosby and Danny Kaye; Metro thought it would make an excellent vehicle for Gene Kelly, and producer William Goetz, working through Columbia, pictured Jane Russell, Betty Grable, and the "King" himself, Clark Gable, in his cinematization. But independent producer Samuel Goldwyn outbid everybody in 1954 when he plunked down a cool one million dollars against ten percent of the movie's worldwide gross. A deal was made with MGM for distribution.

The property acquired, tinsel-town began guessing who would play the show's four leads. Gene Kelly, Cary Grant, Tony Martin, Burt Lancaster, Jeff Chandler, and Dean Martin led the field where the part of Sky Masterson was concerned, while Frank Sinatra, Sam Levene, Danny Thomas, Phil Silvers, and Sid Caesar were the most often mentioned for the Nathan Detroit assignment. On the distaff side, Vivian Blaine, Judy Holliday, and Betty Grable were the likely contenders for Adelaide, and for Sarah the talk was of Rosemary Clooney, Ava Gardner, Ann Blyth, and June Allyson.

As history tells us, Sinatra and Miss Blaine got their sought-after roles, but the part of the Salvation Army lady went to Jean Simmons. The biggest surprise in casting was in the Sky Masterson role. For the handsome gambler, Goldwyn chose nonmusical method actor Marlon Brando.

To direct and write the screenplay for his musical, which would have a production budget of four million dollars *plus,* Goldwyn decided to use a director who had never before done a musical. Joseph L. Manchiewicz owned four Oscars and was responsible for such fine pictures as *All About Eve,* but musicals for him were virgin territory.

Michael Kidd was imported from the Broadway company to repeat as choreographer, as were Stubby Kaye, B.S. Pully, and Johnny Silver to again play Nicely-Nicely, Big Jule of Chicago, and Benny Southstreet respectively.

Loesser added new songs ("Pet Me, Poppa,"

Guys and Dolls. **Jean Simmons and Marlon Brando.**

"Adelaide," and "A Woman in Love"), and dropped others ("My Time of Day," "I've Never Been in Love Before," "A Bushel and a Peck," and "More I Cannot Wish You"). On stage Nathan had not done much singing, but with Sinatra playing the part that, of course, was changed.

"In the play," Manckiewicz told *Cue,* "Sky joins the Save-A-Soul Mission Army. That didn't seem to me logical for a movie musical. Now, we have a big double wedding. Adelaide finally bags Nathan Detroit after their 14-year engagement, Sky marries Sarah, and everybody's happy. We know Sky is going to be a good husband and father, but he won't be a brass-buttoned, uniformed Save-A-Soul-er. He's not the type. It just wouldn't happen.

"But we haven't changed anything essential."

The director found there were certain problems in adapting Runyon to the screen: "Many writers have tried to reproduce Runyonese, but it's just about impossible. There's a lilt and rhythm that baffles you. In the Havana sequence, which I had to rewrite some, I have Sky's speech closer to normal and less like Runyon—basically because Runyonese isn't an

Guys and Dolls. **Jean Simmons and Marlon Brando in the "If I Were a Bell" number.**

Guys and Dolls. **Frank Sinatra and Vivian Blaine.**

Guys and Dolls. B. S. Pully, Sheldon Leonard, Johnny Silver, Dan Dayton, and Frank Sinatra watch as Marlon Brando rolls the dice and sings "Luck Be a Lady."

exportable product; it can't be transported away from Broadway."

Goldwyn's *Guys and Dolls* had all the fun and vitality of the stage original. Rather than try for realism, Manckiewicz wisely chose to play his Broadway in a stylized manner—with many of the billboards and buildings colorfully painted on a cyclorama. Kidd's staging of the musical numbers—such as the choreography to the opening "Runyonland Music" and, later, the title song—was superb. And Loesser's new songs, especially the mellow "A Woman in Love," were also fine.

Vivian Blaine gave a capital performance in the recreation of her stage role, as did Stubby Kaye with his "Sit Down, You're Rockin' the Boat." Brando and particularly Miss Simmons were more than competent in their musical debuts. Only Sinatra seemed less than perfect. Although he was certainly

Guys and Dolls. Johnny Silver, Frank Sinatra, and Robert Keith listen to Stubby Kaye sing "Sit Down You're Rocking the Boat."

okay, the role of Nathan might have been better executed by someone with more pronounced comedic tendencies.

Critical comment on *Guys and Dolls* was generally enthusiastic. *Time* said it was "a 158-minute blur of unmitigated energy, and one of the year's best musicals."

Guys and Dolls grossed approximately eight million in domestic film rentals, making it second only to *The Best Years of Our Lives* as the most financially successful film Goldwyn ever produced. Like the movie version of *Oklahoma!*, it was a fine tribute to its legitimate origins and to the man who started it all—Damon Runyon.

7

Carousel

(1956)

Flush with their success of *Oklahoma!*, Rodgers and Hammerstein signed with the Theatre Guild to create another show. This time they would use their revolutionary new approach to musicals (book dictates technique) on a more serious story than Lynn Rigg's folk play. The adapted work would be Ferenc Molnar's moody fantasy, *Liliom*, which had been successfully produced by the Guild in 1921.

The collaborators switched the play's setting from Hungary to New England during the 1870s. The Americanization provided their story with a colorful background full of local customs and speech, and also ensembles of sailors, fisherman, and mill girls. The leading character remained a tough good looking amusement park barker, but his name was changed from Liliom to Billy Bigelow. This alteration made necessary a new title for the show—*Carousel*.

The basic story line can be quickly summarized. Julie, a sweet young girl, falls in love with the likeable ne'er-do-well, Billy, marries him, and then loses him when he is accidentally killed in an abortive robbery, trying to get money for her and an expected child. Fifteen years later, Billy returns to earth to straighten out his unhappy, maladjusted daughter, Louise, and thereby make amends for the misery he caused Julie.

Carousel opened in New York in April of 1945 at the Majestic Theater. As with *Oklahoma!*, Rouben Mamoulian directed and Agnes deMille choreographed, including in her contribution a dramatic ballet sequence that was similar in tone to the dream ballet in *Oklahoma!* John Raitt and Jan Clayton starred in the production.

Rodgers and Hammerstein again broke new ground with this musical venture. The play opened in the amusement park to the strains of "The Carousel Waltz," a haunting, symphonic melody that continued through the entire first seen. Nothing was spoken or sung during this involved prelude, yet the major characters were established, the initial conflict set in motion.

Little of the music in *Carousel* possessed the light quality present in the creators' earlier work. Indeed, songs like "Soliloquy," in which Billy pondered the responsibilities of approaching parenthood, and the inspiring and now classic, "You'll Never Walk Alone," were interwoven with the emotional fabric of the story. Brief reprises and musical interludes, played under the dialogue, also underscored and added to the dramatic effectiveness. "If I Loved You," possibly Rodgers and Hammerstein's finest love ballad, and "June Is Bustin' Out All Over" were two other numbers that prompted critics to

Carousel. **Audrey Christie, Gordon MacRae, Shirley Jones, and Barbara Ruick.**

Carousel. **Gordon MacRae and Gene Lockhart.**

praise *Carousel*. John Chapman of the *New York Daily News* called it "one of the finest musical plays I have ever seen," while Brooks Atkinson writing in the *New York Times* of the 1954 Broadway revival, said, "This is the most glorious of the Rodgers and Hammerstein works. When the highest judge of all hands down the ultimate verdict, it is this column's opinion that *Carousel* will be the finest of their creations. "

The original production of *Carousel* ran for 890 performances, and was voted both the New York Drama Critics and Donaldson Awards as best musical of the season.

A decade passed before Darryl F. Zanuck, production chief at Twentieth Century-Fox, negotiated a deal to turn *Carousel* into a motion picture. The plan was to star Frank Sinatra as Billy Bigelow, and film the musical in a new wide-screen process, CinemaScope 55, for a reported budget of five million dollars. Unlike *Oklahoma!*, Rodgers and

Hammerstein were not directly involved in the picture's production.

Producer Henry Ephron wrote the screenplay for *Carousel* with his wife, Phoebe. Henry King directed, and Rod Alexander adapted Agnes deMille's original choreography to the movie version. A highlight of Alexander's work was the lively out door staging of "June Is Bustin' Out All Over" on a fishing wharf.

Shirley Jones, who'd just completed the yet unreleased *Oklahoma!*, was cast as Julie, while other key roles were taken by Cameron Mitchell as Billy's nefarious pal, Jigger; Barbara Ruick as Julie's friend, Carrie, who delivered two pleasant tunes, "When I Marry Mister Snow" and "You're a Queer One, Julie Jordan"; and Robert Rounseville as Mister Snow.

Location filming had barely begun in Boothbay Harbor, Maine, when Frank Sinatra shocked both the company and Hollywood by quitting the production. The trade was baffled by his move because Billy Bigelow was a choice role—one that any number of performers had begged to do.

Sinatra justified his action by claiming he would not do two pictures for the price of one. Since the film was being shot in both 55mm and regular CinemaScope, actors were required to perform each scene twice. "The Voice" reportedly said that he couldn't do justice to the picture or himself if he had to do such double duty. Besides, from a legal standpoint, it seems the performer had never gotten around to affixing his signature to a contract.

(Ironically, Fox soon dropped the dual filming when it was discovered that the 35mm CinemaScope version could be achieved through a comparatively inexpensive laboratory reduction process.)

Some artists on the *Carousel* location would later theorize—in private—that Sinatra used the double filming excuse to mask his true reason for quitting. They claim that once on the set Frank became insecure with his ability to play Billy properly and therefore sought a way out.

Whatever the reason, Fox had an expensive crew of performers and craftsmen sitting around in Maine sans a leading man. A replacement was needed—*fast*.

The two most obvious choices were Howard Keel, John Raitt's replacement in the New York production, and Gordon MacRae, who'd recently played Billy at the Dallas State Fair. Keel was under contract to MGM, which could mean a complicated loan-out negotiation, while MacRae was currently freelancing. He therefore got the nod and, following a commitment to play a weekend at the California State Fair in Sacramento, flew to Maine for three weeks of location work.

"I thought the way they shot *Carousel* was stupid," states MacRae. "The whole thing was a flashback so that, when the film begins, Billy is already dead. If I'd been in on the movie from the start, I'd have argued that they should have played it without flashbacks . . . as in the play. Frankly, I think that Zanuck was afraid of doing a fantasy picture and played away from it as much as possible."

Although it may have necessitated the cutting of one number—Billy's "The Highest Judge of All"—the idea of beginning the picture with Bigelow in heaven, remembering his life on earth, was not structurally damaging. The 1956 film remained true to the Hammerstein text, allowing all the emotion and sentimentality to come through. If anything, the flashback device made audiences curious as to how Billy got where he was, and prevented any unsophisticated viewers from thinking the movie was over when the hero was killed two-thirds of the way through the picture.

Director King may not have had Oscar Hammerstein's watchful eye upon him while he was making *Carousel*, but this didn't encourage him to use the film medium much more than Fred Zinnemann had in *Oklahoma!* Seldom did he veer from the original staging. He wisely allowed the songs and basic text to carry the picture. He was at his most creative in his use of the authentic East Coast scenery and the California locales that doubled for Maine. Here his camera was quite fluid. But, because of New England weather problems, the outdoor shots were inter-cut with "exterior" scenes unmistakably filmed on soundstages, and some of these abrupt transitions were jarring—though not overly so.

This technicality aside, *Carousel*, like *Oklahoma!*, was a marvelous production, bringing to a mass audience Rodgers and Hammerstein's stirring achievement. MacRae and Miss Jones were excellent, singing and acting roles that offered much more depth than Curly and Laurey. MacRae's rendition of "Soliloquy" as he strolled along the beach, surf pounding against the rocks, and the final chorus offering of "You'll Never Walk Alone" at Louise's graduation ceremony were just two grand moments that made the film a memorable experience.

Carousel. Shirley Jones, Gordon MacRae, Richard Deacon, and John Dehner.

The *New York Times* called it "a beautifully turned out film, crisply played and richly sung by a fine cast that is fully worthy of the original musical show. . . . Seldom has a musical comedy been made to look more handsome on the screen."

Carousel was completely ignored in the 1956 Academy Awards race. Choosing not to split its studio votes, Fox pushed another of its Rodgers and Hammerstein projects, *The King and I,* which was released later that same year. Strategically it turned out to be the right move, since the Deborah Kerr/Yul Brynner starrer won a number of awards, including best actor. The decision by the studio brass may have been unfair to the artists involved in the earlier picture, but then that's how the Oscar game is played.

Carousel. Gordon MacRae and Audrey Christie.

Carousel. Robert Rounseville, Shirley Jones, Gordon MacRae, Barbara Ruick, and Cameron Mitchell.

Carousel. Jacques D'Amboise in the ballet sequence.

8
Pal joey
(1957)

Pal Joey was perhaps Broadway's first adult musical, and as such was way ahead of its time when it debuted back in 1940. Its more or less realistic characters and situations were unlike those found in the saccharin musical comedies that were then being seen on New York stages. The "hero" of this play was a "gilt-edged heel," who'd sprung from the seamy, sex-filled world of novelist John O'Hara.

O'Hara's Joey Evans first saw the light of day in a series of stories published in *The New Yorker*. The author himself ultimately came up with the idea to turn these pieces into a musical. He approached the composing team of Richard Rodgers and Lorenz Hart, who agreed to collaborate. O'Hara, of course, would do the book himself.

This musical comedy with a mature theme opened on Christmas Day, 1940, at the Ethel Barrymore Theater. It was produced and directed by George Abbott, choreographed by Robert Alton, and gave Gene Kelly the role that made him a star.

Set in Chicago, *Pal joey* tells of Joey Evans, a nightclub hoofer who will stop at almost nothing to get ahead, either in show biz or with the ladies. His well-tuned charm attracts the attentions of Gladys (June Havoc), a member of the chorus line at the small club where he works, and Linda (Leila Ernst), a *nice* girl who clerks in a pet shop. Then there is Vera Simpson (Vivienne Segal), the rich lady from the Gold Coast, who becomes enamoured with this cad, sets him up in a fancy apartment, and buys him his own *club—Chez joey.*

Blackmailers plan to reveal to Vera's husband her affair with Joey, but, thanks to a warning from Linda, the matron thwarts their efforts. Vera tires of Joey's seeing other women, like Linda, and sends him packing. Evans leaves town alone, knowing he will find another comfortable situation elsewhere.

Pal joey, which, unlike most musicals of that day, explored its three-dimensional characters with considerable depth through both song and dialogue, contains two of Rodgers and Hart's most endearing numbers, "I Could Write a Book" and "Bewitched, Bothered and Bewildered," as well as such clever and suggestive tunes as "Zip" and "In Our Little Den of Iniquity."

"Mr. O'Hara, though new at this business, has been able to sustain some semblance of intelligence in his book," said Burns Mantle in the *New York Daily News*. "George Abbott, the producer, has picked an excellent company, the Messrs. Rodgers and Hart have splattered the score with pleasant songs and Robert Alton has contributed his usual assortment of arresting dances . . . . *Pal Joey* adds definitely to the current competition in musical

Pal Joey. **Frank Sinatra arid Kim Novak.**

Pal Joey. **Frank Sinatra and Rita Hayworth.**

entertainment, and this is cheering."

Unfortunately, most critics didn't go along with Mantle's praise of this daring show. The general consensus—from both reviewers and the public—was that the story was too distasteful and, whatever the play's other merits might be, these were not enough to compensate for the underlying unpleasantness. 374 performances after opening, the production closed.

In 1952, *Pal Joey* was revived, with a cast that featured Harold Lang in the title role and Vivienne Segal playing Vera again. This time the reception was different. Neither critics nor the public were shocked. Everyone was raving about the concise writing and versatile score in this exciting show, which set a record for musical revivals by running a total of 540 performances.

Pal Joey. **Frank Sinatra, Barbara Nichols, and Kim Novak.**

Pal Joey. **Rita Hayworth sings "Bewitched, Bothered, and Bewildered."**

Columbia Pictures had purchased the movie rights to *Pal Joey* back in 1941, but for reasons of casting (Cary Grant and James Cagney were mentioned as potential Joeys), script, and, in particular, censorship, production on the property was continually postponed. 1951 found the project reactivated—the rumor being that Gene Kelly, by then a major star at Metro, would be borrowed from that studio to appear in the *Joey* film adaptation with Columbia's sex-goddess, Rita Hayworth. The pair had previously co-starred in Columbia's highly successful *Cover Girl.*

But it wasn't until 1957 that the Rodgers and Hart musical found its way into movie houses. Produced by Fred Kohlmar for Columbia, scripted by Dorothy Kingsley, and directed by George Sidney, the Technicolor release had Miss Hayworth top billed as Vera, and her successor at the studio, Kim Novak, as Linda. Joey was no longer a dancer but a singer—the change being made because Frank Sinatra had been cast in the part.

Filmmakers in 1957 still had to contend with the Production Code and, as a result, *Pal Joey* suffered the censor's blue pencil. For example, all but five songs from the original show were cut—"In Our Little Den of Iniquity" being one of the victims. To make the surviving numbers more acceptable, lyrics were softened as deemed necessary, or, as in the case of "Zip," the number was utilized in an entirely new context. On stage, "Zip" was done as a mocking strip tease by a girl reporter, but since this character had been dropped from the film the song was now given to Miss Hayworth.

To compensate for the missing songs, the producer and director substituted four other favorites by the composers—"My Funny Valentine," "I Didn't Know What Time It Was," "There's a Small Hotel," and "The Lady is a Tramp."

Interestingly, *Pal Joey* was not an integrated musical film, but a play with music. With the exception of "Bewitched, Bothered and Bewildered," which Vera sings to herself as she rises and showers one morning, and the aforementioned "Zip," sung at a charity ball, all of the show's numbers were crooned by Sinatra in the nightclub or in a similar context. In other words, there were no big production numbers and nobody broke into song as he strolled down the street.

Censorship also toned down the sexual innuendoes in the situations and dialogue, yet, as George Sidney told the press, "We haven't softened Joey; he is like John O'Hara wrote him. The guy was a real character in a lilacs-and-honey era, broad-minded in all senses. To him, every girl is still a 'mouse,' but he also uses modern jive talk: everything good is 'a gasser.'

"In fact, it's hard to know where Joey leaves off and Frankie begins."

Producer Kohlmar explained the story changes with "It resolves down to whether we do it in good taste or not."

Other alterations: The story locale was switched from Prohibition era Chicago to modern San Francisco; the role of Gladys, as portrayed by Barbara Nichols, was trimmed down to minor status; the blackmail subplot was eliminated; Vera is no longer married, but a rich widow; and Linda doesn't work in a pet shop. She is a *nice* girl in the chorus line who, thanks to Vera's playing Cupid, strolls off into the sunset with Joey at the picture's conclusion.

Frank Sinatra was perfectly cast as Joey Evans, playing this unsavory character with a chip on his shoulder, yet with just enough charm to make him lovable to the audience. One of the crooner's best moments was his singing of "The Lady Is a Tramp" to Vera when she came to see him at the nightclub. And Rita Hayworth was every bit as good as the world-weary, sharp and sexy Vera. Indeed the whole production, despite the cleansing, was strong-with simple, crisp direction supplied by Sidney and a deft screenplay by Miss Kingsley that retained O'Hara's original intent.

Dick Williams of the *Los Angeles Mirror* wrote, "Enough spice and ribald realism remains in this account of a nightclub lady-killer and the rich widow he mesmerizes to make *Pal joey* one of Hollywood's raciest movies of the season. Also, I suspect, one of its most popular."

Pal Joey grossed $4.7 million in domestic film rentals. It has remained a popular entry on television and at movie revival houses, and is an excellent example of how a film musical can succeed without resorting to a series of dazzling production numbers that feature a screen full of dancers and special effects.

Pal Joey. **Kim Novak and Frank Sinatra in the "What Is a Man?" dream sequence.**

9

Damn Yankees

(1958)

The Faust legend has been dramatized again and again—from plays by Marlowe and Goethe to *All That Money Can Buy,* the 1941 movie version of Stephen Vincent Benet's *The Devil and Daniel Webster,* to that forgettable Richard Burton/Elizabeth Taylor starrer, *Hammersmith ls Out* (1972). It's even been done as a successful stage musical *comedy: Damn Yankees,* produced in 1955, and featuring music and lyrics by Richard Adler and Jerry Ross of *The Pajama Game* fame.

Adapted by Douglas Wallop and George Abbot from Wallop's novel, *The Year the Yankees Lost the Pennant, Damn Yankees* gave audiences an evening of harmless, lively fun. It ran for 1,019 performances. Abbott directed and Bob Fosse choreographed a cast featuring Stephen Douglass, Ray Walston, and that electrifying dancer *I* comedienne, Gwen Verdon.

Briefly, the amusing plot tells of Joe Boyd, a middle-aged realtor and avid baseball nut who is distraught because his team, the Washington Senators, can never seem to win a game. The devil, known here as Mr. Applegate (Walston), offers him the chance to become the Senators' star player, winning for the team both the pennant and the World Series. His price, of course, is Joe's soul.

Joe agrees-leaving himself an "out clause," and Applegate turns him into Joe Hardy (Douglass), a young, strapping athlete who joins the team and starts winning games.

Disturbed that Joe keeps thinking of the wife he left back home, Applegate calls in Lola (Verdon), a beautiful 172-year-old witch. Her job is to keep Joe from yearning for his past life.

So that the Senators will lose a key game, · the devil, who is secretly rooting for the Yankees, changes Hardy back to Boyd just as he is about to catch a pop fly. The tired Boyd rushes home to his wife, wishing to forget about his supernatural experience. Applegate is unable to lure him back.

"You've Got to Have Heart" and "Whatever Lola Wants" were the two major song hits to emerge from *Damn Yankees.* But there were also a number of other nifty tunes in the Adler/Ross score : "Six Months Out of Every Year," a baseball widow's lament; "The Game," a slightly bawdy piece sung by the ball players; "Those Were the Good Old Days," in which the Devil reminisces; a hoedown, "Shoeless Joe from Hannibal, Mo.," and a mambo, "Who's Got the Pain?" Wallop and Abbott's book was successful in finding an entertaining new treatment for a familiar theme. Lines were clever, the pace was brisk, and performances, particularly those of Walston and Miss Verdon, sparkled.

Damn Yankess. Gwen Verdon.

Damn Yankees. Tab Hunter, Gwen Verdon, and Ray Walston.

Damn Yankees. Gwen Verdon, Tab Hunter, and Ray Walston.

Damn Yankees. "Who's Got the Pain," performed by Gwen Verdon and Bob Fosse.

Writing in *Theatre Arts,* Maurice Zolotow praised the production, especially Abbott's directorial contribution: "A show like *Damn Yankees* has about it the fascination of a fine Byzantine mosaic. At a distance it is a gaudy pageant. Regarded closely, it becomes an artfully assembled design in which many small pieces have been fitted together by a master craftsman."

Damn Yankees won the Antoinette Perry Award as best musical of the season. Gwen Verdon and Ray Walston won as top musical performers, Russ Brown, who played the team manager, was voted best supporting actor in a musical, and Fosse got the Tony for his imaginative choreography.

When the brothers Warner purchased movie rights to *Yankees,* the studio brought to Hollywood virtually the show's entire production team, headed by George Abbott, who would write the screenplay and, with Stanley Donen, produce and direct. In the directorial department, Abbott essentially staged the play, while Donen was in charge of camera matters. Again, Bob Fosse was there to choreograph.

Abbott wanted his Broadway cast to repeat their roles in the picture, but Warner Brothers and good sense dictated that at least one movie name would be necessary to insure the picture's box office success. Somebody suggested that Marilyn Monroe should play Lola. "We had lunch with her one day," remembers Abbott. "I doubt if she was really interested in the part. She was just pulling our legs."

Adds composer Richard Adler, "I don't believe Marilyn could have done the dancing the role required."

Ultimately, Gwen Verdon, then appearing on Broadway in *New Girl in Town,* was signed to reprise Lola, and Walston, performing in *Who Was That Lady?,* was set for Applegate. Both performers had to obtain temporary leaves from their respective shows in order to accommodate the studio. Among the other original cast members signed were Shannon Bolin (Mrs. Boyd), Robert Shafer (Boyd), Russ Brown, Elizabeth Howell, Rae Allen, and Jean Stapleton, the future Edith Bunker.

And who was the "movie name"?

Abbott: "Warners put a lot of pressure on us to cast Tab Hunter, who was one of their contract players. It wasn't the best idea in the world, but we went along with it.

"Hollywood studios sometimes come up with unusual thoughts for casting. When we did *The Pajama Game* on film, they tried to get me to use Bing Crosby in the John Raitt role. He'd have been all wrong."

Adler: "On the surface, Tab Hunter as Joe Hardy was not such a bad suggestion. He had a face that was so handsome that it didn't look real. He could have been a composite man created by God . . . or the devil."

As Hardy, the naive young baseball superman, Hunter's acting performance was fine, but his thin singing voice left something to be desired. Songs like the sweetly sentimental "Goodbye, Old Girl" and "Two Lost Souls," an ensemble number with Verdon, suffered because of his vocal inadequacies.

Devotees of the stage production were disappointed to find one of the more popular tunes, "The

Damn Yankees. **Ray Walston, Robert Shafer, and Shannon Bolin.**

Game," missing from the 1958 movie. That locker room quartet number was too suggestive to get by the censors. Also gone was "A Man Doesn't Know," replaced by a new and inferior Adler/Ross song, "The Empty Chair." Explains Adler, "Nobody really liked 'A Man Doesn't Know'—not Abbott, Fosse, or even myself. I felt it had become too cliched since we did the play. I was asked to write another song, and I readily agreed."

Aside from some jabs at Hunter and a comment from *Newsweek* that "the laughs are infrequent," the Technicolor production was generally well received by critics. It was a diverting entertainment that took no special effort for audiences to enjoy. Though Walston and Miss Verdon both played their assignments in a broad stage technique, it made no difference, since they received the lion's share of the good notices. For Verdon in particular the movie was an excellent vehicle. Insiders claimed that she could look forward to a brilliant new career in films. Said *Variety,* "Her eccentric dancing and singing are stylishly engrossing, and her zany comedy adds to the screen one of its finest new comediennes."

The movie career didn't happen, unfortunately, and Gwen Verdon went back to Broadway, where she later starred in such hits as *Redhead, Sweet Charity,* and *Chicago.*

"I wasn't too happy with the movie," reports Adler. "It stayed too close to the play and didn't utilize the medium of film enough."

Abbot admits that "the film could have been better. We tried different bits—like having the devil get dressed to a speeded-up camera—but many of these didn't work out."

Damn Yankees may not have been the perfect film, but does a film have to be flawless for an audience to leave the theater with a smile on its face?

10

Porgy and Bess

(1959)

Playwrights Dorothy and DuBose Heyward were in a seemingly enviable position back in 1933. The Theatre Guild wanted them to adapt their 1927 play, *Porgy,* into a musical comedy—with a score by Jerome Kern, and the one-and-only Al Jolson as star.

But DuBose had other ideas for his saga of Negro life on Catfish Row. Several years earlier—before he and his wife had written the dramatization of his novel, *Porgy*—George Gershwin had expressed a desire to turn this meaningful work into an opera. Other commitments prevented Gershwin from doing so at that time, yet now, if the composer was still interested, Heyward was willing to refuse the Theatre Guild's idea in favor of this more artistic adaptation.

Gershwin was still enthusiastic. He began working on the project almost immediately and even visited Charleston, South Carolina, the story's locale, to get a feeling of the Negro idiom and tenement life.

The folk opera was completed in September 1935, although Gershwin continued to make changes into the rehearsal period. His brother, Ira, wrote the lyrics to the score with DuBose Heyward, who was credited solely with the libretto. Following a successful Boston tryout, the composer's masterpiece opened in New York at the Alvin Theater in October of 1935. Rouben Mamoulian directed the atmospheric production for the Theatre Guild. He'd also staged *Porgy* back in 1927.

Porgy and Bess is a love story. Porgy, played in the original production by Todd Duncan, is a cripple who lives with his goat in the deprived Negro area of Catfish Row. Bess (Anne Brown), the flashy woman of local bully Crown (Warren Coleman), is deserted by her boy friend when he flees after killing a man. Shunned by the tenement residents, she turns down the advances of a pimp/drug pusher, Sportin' Life (John W. Bubbles), in favor of Porgy. The couple live idyllically until Crown returns for Bess. As a hurricane rages about them, Porgy fights and kills Crown, then later is taken to jail for questioning. His lady, thinking .her man is. gone for good, leaves for New York with Sportin' Life. When Porgy is released, he sets off with his goat for the big city, vowing to find his Bess and bring her home.

Rich with songs that captured the sound and spirit of the southern Negro and his folk music in the early years of this century, Gershwin's magnificent score included several pieces that have become classics of

Porgy and Bess. **Sidney Poitier and Dorothy Dandridge.**

the first caliber. "Summertime," "I Got Plenty o' Nuttin'," "Bess, You Is My Woman Now," and "It Ain't Necessarily So," are melodies that have achieved a viable existence beyond the show's text, while' other fine songs—"A Woman Is a Sometime Thing," "A Red-Headed Woman," "There's a Boat That's Leavin' Soon for New York," and "I'm On My Way"—provided strong moments in performance.

Perhaps the deft mix of popular, jazz, and classical musical elements into what was the first truly American folk opera was just too revolutionary for critics and audiences to appreciate back in 1935. After all, we were still several years away from the more lightweight *Oklahoma!* Or maybe it was Mamoulian's artistic staging that turned people off. At any rate, reviewers blasted *Porgy and Bess* in its initial New York production, citing in certain instances Heyward's sometimes awkward plot manipulations. The show closed after 124 performances.

1942—nearly a half-decade after Gershwin's death—saw a revival of *Porgy and Bess.* Cheryl Crawford produced that version which, unlike its predecessor, was a remarkable success, playing 286 performances before it closed. The critics had finally recognized *Porgy and Bess* for what it was—a very important contribution to the lyric theater.

Soon the opera was being played to appreciative audiences around the globe. An acclaimed 1952 tour featured Cab Calloway as Sportin' Life, a role that the bandleader made his own.

When Samuel Goldwyn agreed to pay $650,000 plus a percentage of the gross receipts for the picture rights to *Porgy and Bess,* there was no way he could foresee that this production would be one of the most troubled projects of his lengthy career. Almost immediately, he hired Rouben Mamoulian to direct and N. Richard Nash to write a screenplay.

The next question the filmmaker had to face was casting. Black leading men of star stature were a rare commodity in Hollywood of the late fifties. Harry Belafonte was best known, but he wasn't interested in playing Porgy, so Goldwyn focused his attention on a "comer" named Sidney Poitier.

A few days after the deal with the actor was made, Poitier changed his mind and backed out of the assignment because, according to Goldwyn, he wasn't given script approval. Goldwyn claimed that

Porgy and Bess. **Sidney Poitier.**

Porgy and Bess. **Diahann Carroll.**

he had never given script approval to any performer.

What concerned Poitier was how the Negro characters would be presented in the film. Back in the twenties and thirties when the non-musical and musical versions of the Heyward work were first staged, old Jim Crow was in his prime. But by 1957 attitudes had changed. Negroes protested the offensive stereotypes that Hollywood and Broadway producers had created, and there was a growing movement afoot to present blacks in a much more favorable and realistic light. The days of Willie Best, Stepin Fetchit, and Butterfly McQueen were over.

Goldwyn was determined to have Poitier as his star, and requested that the actor meet with Mamoulian and himself to discuss the project. "I am happy to say my reservations have been washed away, from a warm and wonderful explanation regarding plans for the picture," said Poitier when he emerged from the conference. He was back on board, and admitted he "was unfair to Mr. Goldwyn in assuming he might mistreat the property. I didn't have to look far to know he has exercised good taste and dignity in the past."

Poitier's singing voice was dubbed in this Technicolor/TODD-AO production, which would be released in 1959 through Columbia, as were those of Dorothy Dandridge, who played Bess, and Diahann Carroll, who, as Clara, opened the play with the singing of "Summertime." The producer felt that neither of these talented ladies possessed the vocal capabilities to do the Gershwin score justice. On the other hand, Pearl Bailey, Brock Peters (Crown), and Sammy Davis, Jr. (Spartin' Life) used their own voices to deliver their musical numbers. Davis, incidentally, had been cast in the film only after Cab Calloway proved unavailable.

Porgy and Bess was to begin filming in early July of 1958. Oliver Smith's sets were completed, Irene Sharaff's costumes ready, the actors and director prepared. But shortly before that start date a disastrous fire swept through the Goldwyn lot, totally destroying the film's sets, costumes, and props. Loss was estimated at $2.5 million. Start of production was delayed approximately eight weeks.

Another bombshell hit the production in late July while its physical components were being rebuilt. Goldwyn fired Mamoulian, citing "artistic differences," and signed Otto Preminger to replace him. Preminger had previously worked with several of the picture's stars when he directed *Carmen Jones* (1954).

"Goldwyn didn't agree with Mamoulian's concept of *Porgy,*" says actor Brock Peters. "I saw the original set before it burned. It was the most dreamlike . . . romanticized thing. Very impressive, but, in hindsight, I don't really know what we'd have done with it.

"The picture became much more realistic under Preminger."

In a 1973 interview with *Encore,* Preminger discussed *Porgy:* "Just before I started to direct the film, I called the whole cast together and said, 'Look, ladies and gentlemen, I have no prejudices against black people. I was born in Vienna. I like or dislike someone as a person. I make *no* generalizations and have *no* racial prejudices. That's why you must be prepared. I will treat you just as well and just as badly as I treat white people and judge you strictly on the basis of your work."

More problems—such as ill-founded charges of racial prejudice against Preminger by a member of the Negro Actors Guild of America, or the director's continuing squabbles with Goldwyn—did not deter production once it got underway. After te.n days of filming exteriors in northern California, the company returned to the studio to shoot the bulk of the seven million dollar movie on a sumptuous soundstage creation of Catfish Row.

Samuel Goldwyn's *Porgy and Bess* was lush, visually attractive, but dramatically ponderous. Other than the sets and costumes, it had only the effective performances of Davis, Peters, and Miss Bailey, plus the Gershwin score, to recommend it.

The basic problem appeared to be Preminger's handling of the project. Whether by his choice or as a result of Goldwyn's orders, the director kept his camera static in full or medium shots for almost the entire movie. Seldom did he move in for a close shot. The only musical number in which the camera appeared to have any movement whatsoever was Sammy Davis's singing of "It Ain't Necessarily So." It was almost as if the viewer was sitting in a theater, watching *Porgy and Bess* on stage— a throwback to the early days of sound when the camera was prevented from moving because it was stuck in a soundproof booth. (Preminger's direction of *Carmen Jones* had also been sterile. In that Fox release, he'd not let his camera participate in the musical numbers either.)

Another problem with the film was probably a result of its creators' going a bit overboard to avoid offending blacks. Said *Time,* "And for some strange, wrong reason . . . the actors speak in precise, cultivated accents that are miles away from the Negro slums of South Carolina. For that matter, Sidney Poitier's Porgy is not the dirty, ragtag beggar of the Heyward script, but a well-scrubbed young romantic hero who is never seen taking a penny from anybody."

Both Miss Dandridge and Poitier delivered sensitive dramatic portrayals, but at times one wished

Porgy and Bess. **"It Ain't Necessarily So," as performed by Sammy Davis, Jr.**

Porgy and Bess. **Dorothy Dandridge and Brock Peters.**

Porgy and Bess. **Pearl Bailey and Sidney Poitier.**

that their rich "ghost" singing voices (Adele Addison and Robert McFerrin respectively) were better matched to the actors' natural vocal capabilities.

"As screen entertainment, *Porgy and Bess* is like opera itself—a sometime thing," said *Variety.* Conversely, the usually overly-critical *New York Times* called the offering "a stunning, exciting and moving film . . . ."

Porgy and Bess, which was the last movie Samuel Goldwyn produced before his retirement, was a box office disaster even though it did win three Academy Awards (scoring of a musical picture, costuming, and sound recording). For the average moviegoer opera was of little interest, and for those who did see it the slow pace was a turn-off.

Still, there was the Gershwin music—beautifully presented. Some reviewers felt that that sole factor made up for a multitude of sins. If the visuals became too trying, one could always lean one's head back, close one's eyes, and just listen. But, then, one could do that with the sound-track album, too.

11
West Side Story
(1961)

When *West Side Story* made its Broadway debut in September of 1957, it was hailed as a major achievement of the American musical theater. Here was a new theatrical form—one in which the element of dance was just as important in telling the story as were music and the spoken word.

Director/choreographer Jerome Robbins and composer Leonard Bernstein had discussed doing a contemporary socially-conscious version of *Romeo and Juliet* as far back as 1949. Their first thought was to call it *East Side Story* and have the plot concern a romance between a Jewish girl and a Catholic boy living on New York's East Side. But they finally decided that such an inter-religious theme was antiquated and abandoned the project.

The idea was renewed several years later when, made aware of the sudden, sometimes violent influx of Puerto Ricans into New York, Robbins and Bernstein decided their basic idea might work better if the elements were changed to a love story between a Puerto Rican girl and an American youth. The setting would switch to the West Side and, for the feuding families of the Montagues and Capulets, two rival teenage gangs would be substituted.

Arthur Laurents signed on to do the book of *West Side Story* and Stephen Sondheim the lyrics. Produced by Robert E. Griffith and Harold Prince, it opened at the Winter Garden Theater with a cast that featured Carol Lawrence, Larry Kert, and Chita Rivera.

It is the Jets, led by Riff, versus the Puerto Rican Sharks, led by Bernardo, that do battle in the slums of New York. When Bernardo's sister Marie (Miss Lawrence), falls in love with ex-Jet Tony (Kert), complications arise. Tony tries to stop a rumble between the two gangs, but when his friend, Riff, is killed by Bernardo, he grabs a knife and slays the Puerto Rican leader. Later, as he runs to meet Maria, Tony is himself murdered—by a member of the Sharks.

The score from *West Side Story* is certainly one of Leonard Bernstein's finest works. Several of the songs have become standards. "Tonight," "Somewhere," and "Maria" are among the most beautiful love ballads ever written. "I Feel Pretty" is a touching melody of a young girl's realization of her womanhood, while both "America" and "Gee, Officer Krupke!" offer sharp, yet amusing, social commentary.

But the factor that made the musical such an exciting evening of theater was Jerome Robbins's choreography. Through his electrifying dances,

79

West Side Story. Russ Tamblyn leads the Jets in the movie's opening sequence.

West Side Story. **Rita Moreno leads the "America" number.**

audiences felt all that was not spoken: the hostility and suspicion of the gangs, the panic of the street rumble, the love between Tony and Maria, and the final despair at Tony's death. If the concept of "total theater" is indeed tenable, then it was definitely present in *West Side Story.* The entire production was one rich blend of drama, music, and dance.

Walter Kerr, in the *New York Herald-Tribune,* said: "Mr. Robbins never runs out of his original explosive life-force. Though the essential images are always the same—two spitting groups of people advancing with bared teeth and clawed fists upon one another—there is fresh excitement in the next debacle, and the next."

West Side Story played 734 performances in its initial New York run.

The hit musical was purchased for the movies by Mirisch Pictures, a company that released its product through United Artists. Ernest Lehman signed on to write the screenplay, and producer Robert Wise was set to co-direct the film with Jerome Robbins. During production, Robbins discussed with the press the problems of transfering his choreography patterns to celluloid: "It's not easy; it

West Side Story. **The dance at the gym.**

West Side Story. **Russ Tamblyn (r) dances at the gym.**

comes down to 'treating' realism. Dancing is not realistic, but much of the action of *West Side Story* is extremely realistic—as realistic as sudden death, with two rival gangs battling for their very existence. The problem, then, was to weld the two, reality and dramatic unreality.

"On the stage we used extremely stylized set tings, which of course were very unreal—settings which would only look silly in a film. This fact called for adapting the choreography to the more realistic settings in the picture."

Although all his basic choreography from the play remained, and he had a strong voice in the movie's editing, Robbins ultimately wound up relinquishing the film's overall directing chores to Wise, who explained to *Variety,* "It was merely a matter of time [not of artistic differences or personal friction]. It took too much time to coordinate our thoughts...."

West Side Story. **Natalie Wood and Richard Beymer.**

West Side Story. **Natalie Wood, George Chakiris, and Richard Beymer.**

It became unweildy and time consuming .... Got to the point where it became unmanageable."

Few basic deviations from the original text took place in the adaptation. The song, "America," was changed to include the Puerto Rican boys as well as the girls, and also to make the lyrics more intelligible; two numbers -"Officer Krupke" and "Cool"—changed positions in deference to the mood, and the dance prologue was doubled in length.

Natalie Wood—singing voice dubbed by Marni Nixon—headed the cast as Maria, Richard Beymer was Tony, Russ Tamblyn played Riff, George Chakiris, who'd done Riff in the London company, was Bernardo, and Rita Moreno took over Chita Rivera's part of Bernardo's girl friend.

From the opening moments of *West Side Story,* when the audience sees the Jets prowling their turf, agile bodies moving with grace and anger in spontaneous bursts of dance, Wise's camera was choreographed with the action. He did not merely photograph it. The viewer was right alongside the youths, feeling their frustration as they danced through the trash-filled slums. "Cool," occuring later in the picture, was fraught with still more tension. The camera tracked backward, leading the on-rushing, finger-snapping Jets as they tried to subdue their mounting rage at the Sharks. The choreography, as on the stage, was the remarkable element in this well-paced movie, but its even greater effectiveness here was due to a large degree to one of the most fluid cameras ever utilized in a musical picture. Special credit is certainly due director of photography Daniel Fapp, and also to Boris Leven for his production design that gave the film's concept its physical being.

All of the six million dollar production's musical sequences were exceptional, with "America," "Officer Krupke," "Tonight," "Cool," and the dance at the gym being most memorable.

West Side Story. **Richard Beymer and Natalie Wood.**

Performances, with one exception, were excellent. Miss Moreno and George Chakiris, in fact, won the supporting actor Oscars for 1961. Natalie Wood was charming as the nubile Puerto Rican girl. Only Richard Beymer, whose voice was also dubbed, seemed uncomfortable in his assignment.

The *Hollywood Reporter:* "*West Side Story* is a magnificent show, a milestone in movie musicals, a box office smash. It is so good that superlatives are superfluous. Let it be noted that the film musical, the one dramatic form that is purely American and purely Hollywood, has never been done better."

West Side Story won a total of ten Academy Awards, including best picture of 1961. Jerome Robbins also received a special statuette for his "brilliant achievements in the art of choreography on film." The movie went on to earn a total domestic rental gross of nearly $19.5 million.

With few exceptions—such as Gene Kelly and Stanley Donen—directorial collaborations in movies are not successful. Nevertheless such a union appears to have worked on screen in the case of *West Side Story*. Each man's talents complemented the other's: Wise, with a special gift for poetic realism, its excitement and inherent natural poignance, and Robbins' innovative ideas in choreography that have become an integral part of virtually every musical made since. A good marriage all around.

12

The Music Man

(1962)

If ever a musical captured the color and spirit of America of yesteryear, it was *The Music Man*. Set in Iowa of 1912, Meredith Willson's play was the Fourth of July . . . Mom's apple pie . . . and the Stars and Stripes, all rolled up into a bouncy, fun-filled piece of nostalgia.

Willson, who was born in Mason City, Iowa, in 1902, was already well established as a composer/conductor on radio when he wrote the book, music and lyrics for his show about a traveling con man. Unfortunately, finding a stage on which to present it was not an easy task, as he explained to the *Los Angeles Times* in 1962: "We couldn't sell it; nobody wanted it; couldn't find backers. In 1955, Rini (Mrs. Willson) and I were doing personal appearances; we were in a little town in Kansas. The phone rang at 1:30 in the morning at the hotel. It was Cy Feuer and Ernie Martin, the producers. They said they'd made a deal with CBS. They'd sold *The Music Man* for $100,000. It was going to be a two-hour special, preempting Ed Sullivan. It would star Ray Bolger and Margaret Truman.

"We were so excited we couldn't sleep. We sat up all night talking about it. Finally, somebody had the faith in the show we had.

"We were on Cloud 9 all week and then Cy and Ernie called again. There'd been an argument with the ad agency about casting. The deal was off. The boat had sailed. You can imagine how low we felt."

Two years later—in *1957—The Music Man* found its way to Broadway. It opened at the Majestic Theater on December 19, under the direction of Morton DaCosta and with choreography by Onna White. The star was Robert Preston, who, frankly, had not been first choice for the assignment—or even second or third. Other more bankable stars had already read the property and, unimpressed with its potential, declined. These performers included Danny Kaye, ·Gene Kelly, and Phil Harris. Preston's name was quite far down the list.

The former Paramount Pictures contract player, making his debut in a musical, essayed the role of "Professor" Harold Hill, a lovable swindler who travels throughout the Midwest convincing folks in small towns that they should form a local boys' band. He has the band instruments and uniforms for sale, and *he* will train the youths. Alas, Hill cannot read a note of music. Once he's collected his money, he disappears.

It's in River City, Iowa-where the entire action of the play is set—that Hill, as he so ably puts it, "gets his foot stuck in the door." After mesmerizing

The Music Man. Buddy Hackett and Robert Preston sing "The Sadder But Wiser Girl for Me."

The Music Man. Robert Preston (c) in the "Seventy-Six Trombones" number.

The Music Man. Robert Preston sings "Marian the Librarian" to Shirley Jones.

The Music Man. Shirley Jones and Robert Preston.

the citizenry into believing their sons' moral character will be saved with a band, he falls for Marian (Barbara Cook), the usually standoffish town librarian. Instead of absconding with the loot, Hill decides to stay, vindicate himself, and marry the girl

The Music Man was stuffed with corn, but it worked. Costuming, sets, dialogue, and, particularly, Willson's songs blended perfectly to present a rich and racy recreation of the period. Opening with the traveling salesmen's number, "You Got to Know the Territory," recited in rhythm to the clacking rail wheels, the play was a mixture of musical styles borrowed from an earlier era. There were barbershop ballads ("Lida Rose") sung by the Buffalo Bills; light social commentary on Iowa manners and customs ("Iowa Stubborn," "Pick-aLittle, Talk-a-Little"); Hill's dazzling spiel, "Trouble"; vivacious dance numbers ("Marian the Librarian," "Shipoopi"); romantic ballads ("Goodnight, My Someone," "Till There Was You," "My White Knight"); and the rousing, show-stopping march, "Seventy-Six Trombones."

The Music Man was the musical hit of 1957, and Preston became one of the hottest actors on Broadway. John Chapman of the *New York Daily News* said the play was "one of the few great musical comedies of the last twenty-six years." It won the Antoinette Perry, New York Drama Critics Circle, and Outer Circle awards-and played a total of 1,375 performances.

"Ever since the Broadway show, I've been looking forward to making the film," Morton DaCosta told the press when he'd been signed to produce and direct the movie adaptation of *The Music Man.* Warner Brothers had purchased the property and planned to make it one of its major releases of 1962. DaCosta continued, "The advantages of exterior shooting, creating a small-town atmosphere, open up whole new possibilities. It isn't enough to recreate a stage show on film."

One of the first questions that had to be settled was casting. Who was going to play "Professor" Harold Hill in the picture? The studio was dead set against Robert Preston, even though Willson and DaCosta wanted him. He was, the Warner brass argued, a New York actor whose name meant nothing at the movie box office.

Either Cary Grant or Frank Sinatra seemed the most probable choices for the role, but, gentlemen that they are, both performers insisted that Preston be given first refusal. Finally, after a concerted effort by both Willson and DaCosta, Warners acquiesced and Preston was signed.

It was a wise decision. Sinatra or Grant might have been perfectly acceptable as Hill, but Preston had already given the part greatness, having made it one of the most memorable performances in recent American theatrical history.

The actor was joined in the movie by Shirley Jones, replacing Barbara Cook as Marian, Buddy Hackett, Hermione Gingold, Paul Ford, and, from the original cast, Pert Kelton and the Buffalo Bills. Marion Hargrove did the screenplay and Onna White repeated as choreographer.

Running 151 minutes, *The Music Man* was a perfect marriage of the stage and screen mediums. Without sacrificing a single value from the original, DaCosta opened up Willson's story, totally abandoning the feeling of a procenium. He moved his camera throughout the town of River City, making this well-detailed, if slightly scrubbed, Hollywood backlot locale very believable. True to his word, he was not satisfied just to shoot the stage show, but seemed to approach the material as if it had been written directly for the screen. Split screen musical numbers and iris dissolves were among the devices he employed to help audiences forget the show's theatrical origins.

The director, from a performance standpoint, captured a quality that is often elusive in film *musicals—spontaneity*. DaCosta's players gave the appearance of singing their numbers right there and then, and did not seem to be merely mouthing words to a playback recording.

Aside from the addition of numerous new set tings, there were only two noticable changes from the original stage text. The song, "My White Knight," was dropped in favor of a new tune, "Being in Love." The motivation for such a move—since both tunes are quite similar—is puzzling until one realizes that, with "Being in Love," the creators had a number that would be eligible for "best original song" come Oscar time.

DaCosta also gave the movie a booming finale. Marching to the tune of "Seventy-Six Trombones," Preston—dressed in a sparkling white bandmaster's uniform—led a parade down Main Street and made audiences wish the film was not over.

Variety tagged the picture "an absolute, unequivocable delight on the screen, probably even better than it was on the stage." *The Music Man* was Oscar-nominated in the categories of best picture, costume design, sound, and art direction. It won the trophy for best scoring of a musical (adaptation).

The movie's domestic box office gross may not have been overwhelming, but it was certainly respectable at $8.1 million. Even more important, however, is the fact that *The Music Man* proved that a movie adaptation can discard the proscenium without destroying the enchantment of the original material.

The Music Man. **Mary Wickes, Sara Seegar, Adnia Rice, Peggy Mondo, and Jesslyn Fax played the ladies of River City.**

The Music Man. Shirley Jones and Robert Preston lead the big parade finale.

13
Gypsy
(1962)

Musical biographies of great entertainers have always been popular with filmmakers. Two of the most successful in the genre—*Funny Girl* (1968) and *Gypsy* (1962)—started as stage productions, both featuring music by Jule Styne. *Funny Girl,* the story of comedienne Fanny Brice, made Barbra Streisand a superstar, while *Gypsy* told the tale of burlesque queen Gypsy Rose Lee and featured a powerhouse performance on stage by Ethel Merman as "Mama Rose." The fact that Miss Merman did not repeat her role in the Warner Brothers adaptation stirred much controversy with audiences and critics.

Based on Miss Lee's autobiography, *Gypsy* opened in New York in 1959 at the Broadway Theater. It was produced by David Merrick and Leland Hayward, directed and choreographed by Jerome Robbins. Book was by Arthur Laurents, and Stephen Sondheim supplied lyrics to Styne's fine score.

Mama Rose is the classic stage mother, a fearsome woman of monomaniacal ambition who is determined to make stars of her two young daughters. She devises a vaudeville act, "Baby June and Her Newsboys," and, with the help of her long suffering manager/boy friend Herbie (played on stage by Jack Klugman), propels the children along the Orpheum circuit during the 1920s. Actually, June, who grows up to become actress June Havoc, and Louise (the future "Gypsy") would like it better if their thrice-married mother would marry Herbie and settle down in a real home. Most of their short lives thus far have been spent in hotel rooms-on a diet of Chinese food.

But Mama will not be swayed from her goal. Even after June runs off with one of the Newsboys to escape her parent's overpowering grasp, Rose tries to develop a new act—featuring the less talented Louise.

Mistakenly booked into a Wichita burlesque house, Rose is about to give in to Louise and Herbie's pleas to quit show business in favor of home and family. Then, one night, the star stripper is arrested and Mama begs her daughter to take the woman's place. "Let's walk away a star," she argues. Louise agrees, but a disgusted Herbie walks out.

The newly christened Gypsy Rose Lee (played on stage by Sandra Church) teases the howling male audience by not taking off any clothes-and she's a hit. They love her. Before long, she's a major star.

Feeling she has lost control and is being shut out of her daughter's life, Mama stands alone on a darkened stage and sings "Rose's Turn," a bitter

Gypsy. Rosalind Russell, Natalie Wood, and Karl Malden.

Gypsy. Paul Wallace and Natalie Wood.

Gypsy. Karl Malden.

parody of Gypsy's act, in which she reveals her true lonely and self-centered psyche. At curtain, Mother and daughter embrace in a reconciliation.

Gypsy was a smash, playing 702 performances in its initial engagement. (A decade or so later, the play would be revived, featuring Angela Lansbury in the lead.) It was also a great personal success for Miss Merman whose portrayal of the play's pivotal character was certainly the show-stopper. "Let Me Entertain You," "Together," "Small World," and "You'll Never Get Away from Me," were the more memorable songs to emerge from the score.

Producer / director Mervyn LeRoy, whose credits included such movie classics as *Little Caesar, The Wizard of Oz,* and *Random Harvest,* saw *Gypsy* in New York and wanted to film it. A close friend of Jack L. Warner, he talked the mogul into buying the property for him. Leonard Spigelgass was signed to write the screenplay, then the business of casting the three million dollar Technicolor production began.

"Ethel Merman is a great talent," says LeRoy, "and I love her. She, of course, was dying to do the picture, but we had to turn her down. No matter how big a star she is on Broadway, her name means very little at the movie box office. *Gypsy* was going to be too important a project to gamble on anything less than a major film star with a proven track record."

Ever since he'd first seen *Gypsy,* LeRoy knew there was only one actress to portray Mama Rose on film, and that was Rosalind Russell. The original Auntie Mame jumped at the chance to play this choice assignment, although she had no illusions about her singing voice. "People come to a musical to hear good music," she said, "but if I'm in it they know they're going to hear damn little music—but they can hope for a good story." Miss Russell's songs were dubbed by Lisa Kirk.

Warner Brothers contract player Natalie Wood grabbed the part of the adult Gypsy, and Oscar winner Karl Malden, whose early show business experience had found him performing in the chorus of a Gilbert and Sullivan troupe, played Herbie. Making an unbilled guest appearance was Jack Benny.

Gypsy. **Natalie Wood and Rosalind Russell.**

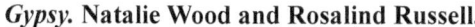

Gypsy. Natalie Wood sings "Let Me Entertain You."

Gypsy. Rosalind Russell sings "Rose's Turn."

In bringing *Gypsy* to the screen, LeRoy decided that he would deviate very little from the New York production. "Why take a great show and improve it into a flop?" he says.

He did, however, open up the film a bit, playing his story on forty-eight sets and in nine especially built theaters. Some minor narration by Miss Russell was added to give the picture more movement, and after some initial previews, the song "Together" was cut because the movie was running too long and this particular number slowed the action.

According to LeRoy, the real Gypsy Rose Lee visited the set quite often: "She cried a few times as she watched her life unfolding."

Gypsy was a moderate box office hit, grossing six million dollars in the domestic market alone. It was a delightful, nicely-paced musical, easily capturing the color and seedy glamour of vaudeville during the Depression. Musical numbers were first-rate, as were the performances of Malden and Miss Wood, who, in maturing Gypsy from an insecure teenager to the beautiful burlesqfie star, gave one of the most moving portrayals of her career.

Only the contribution of Miss Russell received a mixed reaction from critics. The *Hollywood Reporter* said, "She creates a woman whose grimness and tenacity are chilling. She also plays with charm and appeal. This is no anomaly, since determination in itself is not enough to explain these maternal phenomena. Miss Russell's portrayal exposes the mechanics of the personality." On the other hand, Bosley Crowther in the *New York Times* remarked, "she misses the Merman magic and magnificence in the Mama role that is still the big thing in the movie . . . ."

One can easily sympathize with Crowther's purist attitude, which was shared by many. Miss Merman's interpretation of Mama Rose was, no doubt, far more powerful than Miss Russell's, particularly in her rendition of "Rose's Turn." But, then, would Merman have been as effective on film?

It's no secret that many of the most popular stage entertainers have found little success in the motion picture medium. Ethel Merman, Mary Martin, Carol Channing—all have a special "star power" that projects itself across the footlights, yet has eluded capture on celluloid. Merman has made several films, but none of them garnered her a meaningful following in that medium. Even in the movie *Call Me Madam* (1953), in which she effectively recreated her Broadway role, the actress' overwhelming stage technique was too apparent. It didn't really matter in that Twentieth Century—Fox release, because the show was a light, airy musical-a piece of "fluff," not requiring characterization in depth.

Such was not the case with *Gypsy*. Mama Rose is a complex character, requiring the talents of a sensitive actress in order to avoid caricature. Whether Miss Merman could have toned down her brassy personality to achieve the proper effect *on camera* is questionable.

Rosalind Russell brought her performance down to the subtle level of film and, with the aid of Miss Kirk's singing voice, delivered a truly compelling characterization. As much as audiences despised her selfish deeds, they couldn't avoid feeling her pain when, as Herbie leaves, she sits alone and muses, "Why does everyone walk out?"

A reappraisal by the critics of the lady's performance would not be out of line.

14

Bye Bye Birdie

(1963)

Elvis Presley's well-publicized induction into the Army provided the inspiration for *Bye, Bye, Birdie,* a fast-paced satire of the rock-and-roll era. "We wanted to poke fun at the teenagers," says composer Charles Strouse. "The idea was to look at them like insects under a microscope."

With lyrics by Lee Adams and a book by Michael Stewart, *Birdie* opened in April of 1960 at the Martin Beck Theater. Gower Champion made his debut as a stage director with this production and also choreographed.

The story tells of rock idol Conrad Birdie (Dick Gautier), who causes a panic with the nation's teenagers when they learn he's been drafted. Albert Peterson (Dick Van Dyke), Birdie's songwriter/manager, had hoped to marry longtime fiancée Rose Grant (Chita Rivera), but without the commissions earned from his client this will be impossible. Albert's doting, obnoxious mother (Kay Medford) is against her son's marriage and plays on his misplaced feelings of responsibility toward her whenever the wedding date approaches.

Rose hits on an idea that will earn Albert enough money to marry. He will write a new song for Birdie—"One Last Kiss"—to be sung on the Ed Sullivan Show. On that same program, Conrad will kiss one of his many fans goodbye prior to entering the service. The televised kiss—a symbolic gesture for the singer's entire fan following—is expected to boost Albert's song into a multi-million record seller.

The lucky teenager, chosen by chance, is Kim MacAfee (Susan Watson) of Sweet Apple, Ohio. Rose, Albert, and Birdie head for that small town to make all necessary arrangements for the on-air promotion, and, once there, are confronted with some minor complications. Kim has a steady boy friend, Hugo (Michael J. Pollard), who objects to his girl being kissed by Birdie, and her father (Paul Lynde) wants to appear—with his entire family—on the program.

The Sullivan show—occuring at the end of act one—becomes a fiasco when Hugo runs on stage at the moment of the big kiss and coldcocks Birdie. Angry at her steady, Kim goes off with Conrad, but soon becomes disenchanted with his gargantuan ego. She goes back to Hugo, and Albert, who's had a fight with Rose over his mother, finally decides to ignore the nagging old girl and marry the woman he loves.

Bye, Bye, Birdie was a marvelous show, containing many funny lines, some performance gems, wild situations, and a very clever songbook. There was, for example, "The Telephone Hour," in which the

Bye Bye Birdie. Janet Leigh and Dick Van Dyke (c) watch the people of Sweet Apple react to Jesse Pearson's singing of "Honestly Sincere."

Bye Bye Birdie. Dick Van Dyke and Janet Leigh sing "Put on a Happy Face."

Bye Bye Birdie. Ann-Margret sings "How Lovely to Be a Woman."

teenagers of Sweet Apple—each sitting on stage in his or her own cubicle—call each other to discuss the latest school gossip. "Normal American Boy" was a press conference in which Rose and Albert relate diverse accounts of Birdie's background to two different sets of reporters. "Hymn for a Sunday Evening" had the MacAfee family—delightfully led by Paul Lynde-visualizing their upcoming appearance on the Sullivan program. And the comic lament, "Kids," went on to become one of the most popular novelty tunes of the day.

"Spanish Rose" provided a hilarious moment for Chita Rivera when she danced for a group of surprised Shriners. Susan Watson's two big numbers were "One Boy" and "How Lovely to Be a Woman," while Dick Van Dyke scored some engaging points with "Put on a Happy Face" and the finale, "Rosie."

"*Bye, Bye, Birdie* is richly entertaining," was the comment of Richard Watts, Jr. in the *New York Post*. However, Brooks Atkinson in the *New York Times* said, "But as a production, *Bye, Bye Birdie* is neither fish, fowl nor good musical comedy. It needs work."

The general consensus of opinion, fortunately, leaned more toward the sentiments of Watts than that of Atkinson. *Birdie* won the Tony award as best musical of the season, playing 607 performances.

Columbia Pictures purchased the film rights to *Bye, Bye, Birdie* and announced that Gower Champion would direct and Michael Stewart would script. Yet, a few months later, Champion was out—replaced by George Sidney—and, shortly thereafter, Stewart followed. Irving Brecher was now assigned to the screenplay. Reasons for the changes in personnel were never made public.

With the hit stage musical under the control of producer Fred Kohlmar and Sidney, certain shifts in emphasis began to take place in the movie version. "Our play," recalls composer Strouse, "was a satire about teenagers, aimed at an adult audience. But the powers at Columbia wanted to attract a more youthful audience."

To accomplish their goal, the moviemakers enlarged the roles of Kim and Hugo, both of which were secondary in the play. Ann-Margret, newly under contract to the studio and chosen by the producers for a major publicity push, was cast as Kim. Her role now became as large as and, indeed,

Bye Bye Birdie. "Kids," performed by Dick Van Dyke, Maureen Stapleton (as Van Dyke's doting mother), and Paul Lynde.

Bye Bye Birdie. **Bobby Rydell and Ann-Margret dance and sing "A Lot of Livin' to Do."**

nearly overshadowed those of Albert and Rose. When Columbia's *Bye, Bye, Birdie* opened throughout the country, it was the photo of Ann Margret that dominated the ad campaign.

"I thought Ann-Margret was dead wrong for the part," says Strouse. "But, I must admit that she's got a lot of talent and her performance is what made the movie a hit."

Dick Van Dyke and Paul Lynde were both brought from New York to repeat their original roles. Teen singing idol Bobby Rydell was cast as Hugo, and Jesse Pearson, who'd played Birdie in the national company, was called in to do that part on the screen.

Only Rose—the actual "star" role on stage—presented any major problems in casting. Strouse: "There was nobody of true star quality in Hollywood who could do the Chita Rivera role, and the producers didn't want Chita." The part went to Janet Leigh.

Since the teenagers now dominated *Birdie,* the tone of the entire project changed also. Humor became more slapstick and juvenile, rather than satiric. Some of the best numbers from the stage production, including "Normal American Boy" and "Spanish Rose," disappeared. Other songs, like "Put on a Happy Face," which was now enhanced with some animation, were presented in a new context. In sum, most of the songs involving the teens and Birdie were enlarged, while the Rosie/Albert tunes were either altered or eliminated. The two novelty songs, "Kids" and "Hymn for a Sunday Evening," remained essentially the same. Only one song—"Bye Bye Birdie"—was added by Adams and Strouse. It was sung very sexily by Ann Margret prior to the screen credits.

Plot-wise, innumerable new situations and characters were added; relationships were changed. Albert, for example, was no longer Birdie's manager, but only one of his songwriters, and, in Sweet Apple, Rose became briefly involved with a suave high school instructor. Ed Sullivan appeared in several scenes, his program taking place at the movie's climax, rather than at the end of the first act as it had in the stage play. The Sullivan sequence now contained an out-of-key comic sub-plot dealing with Albert's Abbott and Costello-like efforts to speed up a Russian ballet number which had temporarily bumped his song off the air.

Critical commentary on *Bye Bye Birdie,* which earned $6.2 million in domestic film rentals, was, as might be expected, mixed. "Producer Fred Kohlmar has put together an engaging catalogue of commercial merriment, commercial because it's entertaining," was the attitude of *Variety.* The *New York Times* said: "But, unfortunately, Mr. Sidney and his scriptwriter, Irving Brecher, have allowed the essence of this spirited musical comedy of Michael Stewart to get away from them. Not only do they lose Conrad Birdie in the mazes of their rearranged plot, but they lose the essential idea of satire and the pace and sparkle of the show."

In itself, *Bye Bye Birdie* was a well-made, diverting motion picture. The cast was a delight and Sidney moved the story along at a smart clip. Audiences, unfamiliar with the original, must have enjoyed the film immensely. However, for the poor patrons who had seen the award-winning stage show, Columbia's 1963 release must have been a major disappointment. They had certainly expected to view something akin to what they remembered, rather than a big-budgeted cousin to *Beach Blanket Bingo.*

Bye Bye Birdie. **Dick Van Dyke and Janet Leigh sing and dance to "Rosie."**

15
My Fair Lady
(1964)

Time **called it "a delight."** According to Walter Kerr in the New York *Herald Tribune*, it was "miraculous, wise, witty and winning." And Brooks Atkinson of the *New York Times* proclaimed it as "one of the best musicals of the century."

My Fair Lady was adapted from George Bernard Shaw's *Pygmalion*, the story of how snobbish phonetician Professor Henry Higgins turns bedraggled cockney flower girl Eliza Doolittle into a "lady." The great British playwright had for years resisted offers to adapt this work into a musical, although he did allow producer Gabriel Pascal to film a straight motion picture version in 1938. With the exception of having Shaw add an elaborate embassy ball sequence, and also alter the ending so that Eliza wound up with Higgins instead of a rich suitor, little in the movie was different from what had appeared on stage.

Shaw died in 1950 and Pascal in 1954. Alan Jay Lerner and Frederick Loewe became interested in musicalizing *Pygmalion*—after composers like Rodgers and Hammerstein had rejected the idea—and made a deal with the Shaw estate. The creators of *Brigadoon* realized that to make their adaptation work they would really have to do very little to Shaw's script—just include the ball scene and dramatize a few incidents that happened offstage in the original. They also opted to go with the movie's altered ending. Working title for the proposed production was *Lady Liza*.

"Several factors were involved in my decision to do *My Fair Lady*," Rex Harrison told columnist Sheilah Graham in a 1965 interview. The tall, English star had been approached by Lerner and Loewe early in the show's developmental stages. "First, there was Shaw, and the fact that I had guarantees in my contract that certain passages of Shaw would remain. The second factor was genuine faith in the talents of Alan Lerner. He impressed me tremendously and the lyrics he had done were good. I spent about three weeks trying to make up my mind and finally decided to tackle it.

"When it came to the music for me, we were faced with rather a serious limitation—my range. Often we would sit around the piano and all sing Gilbert and Sullivan together. What Loewe was doing was listening to which notes I could sing and which ones I couldn't. It turned out that my range was about two notes, so they did the numbers around those two notes—three at the most."

With Harrison signed for their project, Lerner

My Fair Lady. Audrey Hepburn and Rex Harrison.

My Fair Lady. Wilfred Hyde-White, Stanley Holloway, Rex Harrison, and Audrey Hepburn.

My Fair Lady. Jeremy Brett, Audrey Hepburn, and Rex Harrison.

and Loewe began seeking an Eliza. They wanted Mary Martin, but she was unavailable. After a considerable search, the composers settled on a relatively unknown performer, Julie Andrews, who'd previously starred on Broadway in *The Boy Friend*. Rounding out the cast were Stanley Holloway as Eliza's widower father, Alfred P. Doolittle, and Robert Coote as Higgins's associate, Colonel Pickering.

My Fair Lady opened on March 15, 1956 at the Mark Hellinger Theater. It was directed by Moss Hart and choreographed by Hanya Holm.

The scintillating Lerner/Loewe score was a potpourri of songs, all perfectly attuned to the atmosphere of 1912 London, and all in the spirit of Shaw. "On the Street Where You Live" and "I Could Have Danced All Night" were both runaway hits, while "I've Grown Accustomed to Her Face," "With a Little Bit of Luck," and "Get Me to the Church on Time" have also had a large play over the years.

Attired in Cecil Beaton's exquisite costumes and playing in front of Oliver Smith's felicitious settings, the actors performed some of the most imaginative stagings of musical numbers yet seen on Broadway. The "Ascot Gavotte," done against a black-and-white background, was a sterling moment, as was the renowned "Rain in Spain" number, in which Eliza first accomplishes Higgins's overly tutored basic speech lesson. Then, with a touch of Shavian commentary, there was Higgins's "Why Can't the English?"

Harrison was superb in the musical, endowing his character's plaintive masculine egotism with an irresistible charm. And, as Eliza, Miss Andrews was a total delight. Critics predicted she was destined for major stardom.

When the Antoinette Perry Awards were passed out in 1957, *My Fair Lady* won six, equaling the mark previously set by *South Pacific* (1950) and *Damn Yankees* (1956). The Lerner/Loewe production played 2,717 performances during its initial engagement, and, at this writing, holds the record

My Fair Lady. Audrey Hepburn and Rex Harrison.

My Fair Lady. Rex Harrison.

as the third longest running musical on Broadway, placing after *Fiddler on the Roof and Hello, Dolly!**

Jack L. Warner made news in 1963 when he plunked down $5.5 million for the film rights to *My Fair Lady*. According to the executive, this would be the most expensive picture ever made by Warner Brothers, and also the most expensive movie musical in history, with a projected production cost of another ten million. Warner intended to produce personally.

Casting of the picture, which would have a screenplay by Lerner, became one of the film capital's favorite speculations in the ensuing weeks. Would Rex Harrison repeat the role of Higgins on screen, or would the part go to the more bankable Cary Grant? Would James Cagney agree to break his retirement vow to play Alfred P. Doolittle? And, if not Julie Andrews, who would play Eliza?

By the time veteran George Cukor had signed on to direct, the casting questions were settled. Harrison would, indeed, repeat as Higgins; Cagney refused his offer, so Stanley Holloway was also brought in from the original company; and Wilfred Hyde-White was signed for Pickering. The big controversy arose when Audrey Hepburn, instead of Julie Andrews, was signed to play Eliza. The choice, said Warner, was no reflection against Miss Andrews' abilities. Audrey Hepburn, who received one million dollars and top billing, was simply a more meaningful name to movie audiences—even if her singing voice would eventually have to be dubbed by Marni Nixon.

Transferring *My Fair Lady* to the screen presented George Cukor with a basic problem. He was working with a book that, with minor exceptions, took place indoors and relied heavily on dialogue. It did not really lend itself to wildly innovative cinematic effects. The director, therefore, had to open up his film without going beyond the bounds dictated by the original material. "In the film," Cukor once told an audience at the American Film Institute, "as Eliza and the maids sing 'I Could Have Danced All Night,' Eliza dances up the stairs, and the maids put her to bed. On the stage, the scene had been confined to the set, and she fell asleep on the couch. There can be certain movement without violating the scene or the play."

The "Ascot Gavotte" race sequence was staged by Cukor in a deliberately non-realistic manner. "To shoot it otherwise," said the director, "would have violated the style of the thing. The whole scene was as stylized as 'On the Street Where You Live.' It has its own kind of realism. I grabbed as much as I could from the stage, and that's also a trick: to take something from the stage and not make it stodgy, to make it flow. But this scene had a kind of slightly artificial style. The songs were written that way. We had scenes in Covent Garden, which were more or less realistic, and then the horseracing scene was like a ballet."

My Fair Lady was a sumptuous production, enhanced by Cecil Beaton's master-plan of costumes and scenery, the choreography of Hermes Pan, Harry Stradling's camera, and the masterfully restrained direction of Cukor. Performances seemed faultless, as, indeed, was the entire picture. "It has riches of story, humor, acting and production values far beyond the average," said *Variety*. "It is Hollywood at its best, Jack L. Warner's career capstone and a film that will go on without now foreseeable limits." The often negative *New York Times* also raved about the 1964 Technicolor release, tagging it "a superlative film."

Even though her singing voice had been dubbed, it was difficult to be critical of Miss Hepburn's contribution to the movie. Hers was an endearing portrayal and, with due respect, it is difficult to conceive that Julie Andrews could have given Eliza more. Nevertheless when Academy Award time rolled around, Audrey Hepburn's name was missing from the ranks of best actress nominees, although Miss Andrews did receive recognition—and subsequently won the award itself—for her work in Walt Disney's *Mary Poppins*.

On Oscar night, *My Fair Lady* took home the statuettes for best picture, actor (Harrison), director, costuming, art direction, musical scoring, cinematography, and sound. It went on to become a major moneymaker, earning in excess of thirty million dollars in domestic film rentals.*

Producer/director Norman Jewison gave an interview to the *Hollywood Reporter* in 1971 in which he stated, in part, that *My Fair Lady* was a "bad musical film" because it stuck to the play. "I don't like it when you take something from the stage and put a proscenium around it."

While one can agree in principle with Jewison's thesis, it was certainly his mistake to cite the Lerner/

---

*As this book goes to press, it has been learned via *Variety* that, effective December 8, 1979, *Grease* will play its 3243rd performance, thus surpassing *Fiddler on the Roof* as the longest-running production in Broadway's history.

*This is Warner Brothers' figure. *Variety* claims it only did twelve million in the domestic market.

Loewe play as an example of a *bad* musical movie. True, the Warners release is essentially a filmed version of the stage production, but, then, we must ask Jewison how he would have changed the picture and still come up with such a rich and rewarding finished product.

16
The Sound of Music
(1965)

Until the coming of *Grease* in 1978, *The Sound of Music* was the most popular musical in motion picture history. It was the first movie to dethrone the longtime box office champ, *Gone With the Wind,* and at this writing, with domestic rentals of over $79 million, it ranks sixth in *Variety's* listing of all-time film rental champs.*

Mary Martin conceived the idea of turning the story of the Trapp Family Singers into a musical. She'd seen a German movie about the famous concert hall singers from Salzburg, Austria, and, wishing to play Maria von Trapp, began an eight month search to find the baroness: The hunt ended in an Innsbruck, Austria, hospital, where Maria von Trapp was recovering from malaria. Though reticent at first, the lady finally agreed to the dramatization of her life story after she was informed that the monies she could earn from such a project might be substantial. There were further delays while clearances were obtained from the von Trapp children—then residing all over the globe—but finally Howard Lindsay and Russel Crouse began work on the book, while Rodgers and Hammerstein did the score.

The Sound of Music would be the last collaboration of the composers who gave the world such hits as *Oklahoma!, Carousel, South Pacific,* and *The King and I.* Hammerstein died less than a year after the play opened.

The Lindsay/Crouse script begins in 1938. Maria is a postulant at Austria's Nonnberg Abbey, but the mother abbess and her assistants question the volatile girl's readiness to renounce the outside world. So she will gain perspective, Maria is sent to work as governess to the seven children of wealthy Captain Georg van Trapp. A widower, the Captain rules his household and his children with an iron hand. Maria brings warmth and fun into the youngsters' lives.

After a while, Maria realizes that she loves the Captain, and, upon her return to the abbey, the mother abbess encourages her to pursue this love. The former postulant goes back to the van Trapps.

The Nazis are coming to power in Austria. The Captain, being violently anti-Nazi, breaks off his engagement to fiancée Elsa Schraeder when it is obvious they will never agree on the political issue. Soon his affections turn toward Maria. They are married in the abbey.

The van Trapps get home from their honeymoon to discover that the Nazis are now in power, and the Captain has been summoned to return to his naval

---

*Grease has grossed over $83 million and is ranked fourth after *Star Wars, Jaws,* and *The Godfather.*

The Sound of Music. Julie Andrews, Peggy Wood, Anna Lee, and nuns.

The Sound of Music. Christopher Plummer and Julie Andrews.

The Sound of Music. Eleanor Parker and Christopher Plummer.

The Sound of Music. Julie Andrews and the Von Trapp children.

duties. Family retainer Max Detweeler, however, has a plan. Having been impressed with the singing talents of Maria and the children, he has slated the family to appear at a local music festival. The von Trapps use the concert as a device to elude the brownshirts. After hiding at the abbey, they cross over the mountains to neutral Switzerland.

As with their previous works, Rodgers and Hammerstein gave *The Sound of Music* a score with texture—one that would fit its varying moods and settings. There was the inspirational "Climb Every Mountain," a nod to Austrian folk music with "The Lonely Goatherd" and "Edelweiss," charming children's songs ("So Long, Farewell," and the now-standard "Do Re Mi"), simple, direct character revealing "Maria," "My Favorite Things" and "Sixteen Going on Seventeen," and the unforgettable title song. It was, in the opinion of many, the composers' best score in years.

The charming musical opened in November of 1959 at the Lunt-Fontanne Theater, under the direction of Vincent J. Donehue. Choreography was by Joe Layton. Joining Miss Martin in the cast were Theodore Bikel as the Captain, Kurt Kasznar as Max, Marion Marlowe as Elsa Schraeder, and Patricia Neway as the mother abbess.

Although some reviewers thought *The Sound of Music* was a bit like a dated operetta, most were enchanted by it. John McClain of the New York *Journal-American* said it was "the most mature product of the Rodgers and Hammerstein team. It has style, distinction, grace and persuasion." The production won five Antoinette Perry awards, and played 1,443 performances in its initial Broadway run.

Twentieth Century-Fox purchased film rights to the Rodgers and Hammerstein musical for a reported figure of $1.25 million, and set Oscar winning filmmaker William Wyler to produce and direct. Ernest Lehman was signed to write the screenplay.

For the role of Maria in this nearly eight million dollar film, Wyler cast an actress he'd seen do *My*

The Sound of Music. **Christopher Plummer, Eleanor Parker, and Julie Andrews.**

Fair Lady on Broadway, but who had not yet been seen in a film. She was Julie Andrews, currently shooting *Mary Poppins* for Walt Disney. The Disney people graciously showed Wyler some rushes from their upcoming feature, and he made his decision based on that footage.

After returning from location hunting in Austria, the director was sent a script from Columbia Pictures—*The Collector.* It was a project that excited him much more than the musical he was preparing. Since there was a scheduling conflict between the two films, he requested a release from Fox.

Robert Wise took over the seat of command for the production, which was scheduled to be shot in TODD-AO. Julie Andrews was retained as Maria, while Christopher Plummer was set to play Captain von Trapp. Also cast were Eleanor Parker as the Captain's fiancée, Richard Haydn as Max, and Peggy Wood as the mother abbess.

With eleven weeks of filming to take place in Salzburg, Ernest Lehman opened up the original book to take advantage of the scenic Bavarian Alps.

Scenes were added and, so that the movie would not be overlong, other scenes were scratched. Three songs ("No Way to Stop It," "How Can Love Survive?," and "An Ordinary Couple") were eliminated, and two new ones ("Something Good" and "I Have Confidence in Me") were written for the movie by Richard Rodgers. Two other songs ("The Lonely Goatherd" and "My Favorite Things") were moved around in the picture to occur at different points than they had on the stage.

Lehman also devoted himself to developing a straighter story line. Thus, the roles of Max and Elsa Schraeder (called "The Baroness" in the movie)

The Sound of Music. Angela Cartwright, Julie Andrews, Christopher Plummer, and Charmian Carr.

were reduced considerably. When von Trapp's engagement is broken in the film, it is not for political reasons, but because he realizes that he loves Maria.

According to a press statement by Robert Wise: "Perhaps the biggest noticeable change is in the general mood of the show. From the outset, we—Lehman, associate producer Saul Chaplin, stars Julie Andrews and Christopher Plummer, and I—were all agreed that the more realistic medium of films demanded that the material be treated differently than it was on stage. The sentimentality and gemutlichkeit that worked fine on stage could easily become heavy-handed on the screen.

"To avoid this, there was some subtle tightening up of the dialogue in many scenes and a sharpening of the Captain's character to make him more incisive than he had been on stage. And, both on location and on the sound stages back at the studio, production designer Boris Leven and costume designer Dorothy Jeakins checked constantly and carefully to avoid backgrounds or clothing that might be too gingerbready or cloying."

Richard Rodgers reportedly went along with all of the filmmakers' suggested changes—except where it came to making an alteration in the play's original lyrics. Here, in deference to his late partner, he insisted that Oscar Hammerstein's words remain exactly as written.

No new adjectives can be heaped on the pile of compliments already pllid the film adaptation of *The Sound of Music*. Suffice to say, it was a beautiful picture—full of extraordinary visuals (like that opening shot of Julie Andrews singing in an open field atop the Alps), fine performances and magnificent songs. If there is such a thing, it was the *perfect* adaptation, making full use of the film medium while staying essentially true to the source material. The creators of *West Side Story* had outdone them-selves. *Variety* called it "one of the top musicals to reach the screen."

The Sound of Music has deserved every honor bestowed upon it—including the Academy Award as best picture of 1965.

17
Stop the World–I Want to Get Off
(1966)

Multi-talented Anthony Newley had been a familiar figure in the British cinema and theater since his youth—when he played the Artful Dodger in David Lean's acclaimed film, *Oliver Twist* (1948). His assignments aside from the Dickens classic were rather pedestrian until he and composer Leslie Bricusse got together during the spring of 1961 and, in a ten-week period, wrote an unusual new musical comedy, *Stop the World—I Want to Get Off.*

With Newley both directing and starring, the play opened in London, and, although reviews were mixed, nobody could deny that *World* was endowed with one of the most listenable scores in recent history. Tunes like "Gonna Build a Mountain," "Once in a Lifetime," and, particularly, "What Kind of Fool Am I" were quickly snapped up by dozens of top recording artists. The musical played for over a year.

Producer David Merrick brought the play to New York in October of 1962. Audiences rushed to see the show that had given them such an array of memorable songs. Again, however, the critics were less than enthusiastic. *Newsweek*: "*Stop the World* has its moments—but too few. Newley, costumed up to his neck in baggy clown pants, is in over his head as star, director, and co-author (with Leslie Bricusse) of the book, music, and lyrics."

Set inside a circus tent and utilizing such ingredients as mime, malapropisms, and puns, the musical was, in effect, an Everyman play. Newley, made up in clown white and looking a bit like Marcel Marceau, was "Littlechap," whose life we follow from birth to death. Along the symbolic way he marries the boss's daughter because she's pregnant, has numerous affairs, and advances in position from that of a common laborer to head of the firm, then to Parliament, and finally a peerage.

Backing the action is a Greek chorus of girls, costumed in harlequin-style tights. Littlechap's wife and the three other women (a Russian, a German, and an American) in his life were played by Anna Quayle.

Featuring numbers like "Mumbo Jumbo," a spoof of the British political process, *Stop the World* was a biting satire of life's vicissitudes. It may not have been a perfect play, but it certainly entertained its audiences.

In 1963, Jerome Hellman and Arthur P. Jacobs joined forces to acquire the film rights to the Newley/Bricusse property. The plan was to shoot the picture in New York while the play was still running, and have Newley repeat his chores as director/star. Early in 1964, it was decided that the film might better be made in London. "I've wanted to direct a film for a long time," Newley told the press, "and our show seems ideal for it. . . . At first thought, the idea of filming *Stop the World* might seem incongruous, but it isn't. In treatment, the show is pure mother-earth theater—a raised platform, a small cast used as an audience—in tracing the life of the hero, a selfish man. The very theatrics would seem to be optically appropriate for a movie camera and such effects as a split screen. I'll be doing the scenario, but there'll be very few essential changes."

Unfortunately, Hellman and Jacobs were never able to put a final package together for their project, and they eventually let their option on the material drop.

Bill Sargent had been the first film producer to use a new process from Mark Armistead known as "Electronovision," in which electronic cameras and

Stop the World—I Want to Get Off. **Tony Tanner and players.**

Stop the World—I Want to Get Off. **Tony Tanner and players.**

Stop the World—I Want to Get Off. Tony Tanner and Millicent Martin.

television production techniques were used to achieve a big screen effect. Through this process he'd already brought the disastrous *Harlow* with Carol Lynley and Richard Burton's stage production of *Hamlet* to the screen.

Sargent approached Newley in 1965 with a proposal to film *World* using the Mitchell Camera Corporation's newer System 35, a three-film camera setup with added electronic view-finders permitting monitoring from a console manned by the director. He also wanted the Englishman to both direct and star in the movie, but although Newley was happy to sell Sargent the film rights, he nixed the idea of participating in it. Recalls Sargent, "Tony said the show had done all it could for him as a performer, and by doing the film he would be taking a step backward in his career."

With financing supplied by Warner Brothers, the company that had handled *Hamlet,* Sargent proceeded with his plan to shoot the musical in England on a $637,000 budget.

"Our first choice for the female lead was Millicent Martin," says Sargent, "who British television audiences considered their answer to Lucille Ball. She's a fabulous comedienne. The all-important man's part had us stymied, however, and we tested several actors for it."

Tony Tanner, who would later replace Tommy Steele on Broadway in *Half a Sixpence,* wound up playing Littlechap. (At one point, Donald O'Connor had been mentioned for the role.) Philip Saville directed the Technicolor production, and Michael Lindsay-Hoff was credited as stage director.

A stage director was necessary because Sargent's

Stop the World—I Want to Get Off. Valerie Croft, Millicent Martin, Tony Tanner, and Leila Croft.

Stop the World—I Want to Get Off. Tony Tanner, Leila Croft, Valerie Croft, and Millicent Martin.

adaptation of *World* was little more than a play photographed on a stage. Movie audiences were allowed to view the orchestra, the backstage area, and even the theater audience. Total filming time for the production was approximately two weeks.

Sargent: "There was only one change we made from the stage original. We switched Littlechap's German girl friend to a Japanese. We didn't want to offend the German market."

One nice cinematic device occurs at various points during the movie. When Littlechap reaches a crisis in his life, Tanner holds up his hand, shouting, "Stop the world!" The picture changes to black and-white, and Tanner delivers an introspective aside to the audience.

The performances of Tanner and Miss Martin were definite pluses in the 1966 Warner Brothers release, as was the play's original score. But, as a movie, *Stop the World* just didn't work. The skeleton circus tent setting and surrounding cyclorama

Stop the World—I Want to Get Off. **Tony Tanner, Neil Hawley, and Millicent Martin.**

Stop the World—I Want to Get Off. **Tony Tanner and Millicent Martin.**

were interesting, yet what seemed innovative on the stage came off as contrived on screen. Much spontaneity was lost, the pace was slow, and the device of showing the audience gave the effect of canned laughter. Said Philip K. Scheuer in the *Los Angeles Times,* "even though the camera occasionally comes to life . . . the effect is just about as leaden as when it had to be isolated in a soundproof b'1oth so the mike wouldn't pick up its whirring."

"When they cut the film," says Millicent Martin, "they destroyed our pace and timing. The difficulty of working with multiple cameras is that you play for one shot, but then they use another. For example, in the picture, there's one shot of me looking at the monitors off camera. But, when we did it, I thought they weren't even on me. The editors changed what the director had planned."

"The movie was hurt during editing," agrees Sargent. "Warners rushed it out several months ahead of schedule. They cut twenty-two minutes, including several song reprises and the original ending. Also, the editors didn't use the audience shots properly. An audience is very much a part of live theatre, but in this picture they were utilized only as transition material.

"It did poorly at the box office because Warners booked it immediately into mass audience theaters, rather than art houses where a special film like this could have built a following."

Though the movie adaptation of *Stop the World* was a disappointment to its creators, its failure did not signal any death knell for the filming of live theatrical performances. In 1975, *Give 'Em Hell, Harry,* James Whitmore's one-man show about Harry Truman, was photographed onstage—with Sargent serving as executive producer. It was a well-received production, for which the star was nominated for a best actor Oscar. And, in 1978, Sargent filmed a second version of *Stop the World* starring Sammy Davis, Jr. This adaptation resulted in a lawsuit being filed against the producer by Newley and Bricusse, which, at this writing, has not yet been resolved.

18
Finian's Rainbow
(1968)

Brigadoon and *Finian's Rainbow* have much in common. Both musicals debuted in 1947. Both were smash box office hits. And, both, though distinctive in their own ways, dealt with imaginary legends—one Scottish and the other Irish.

A Lerner/Loewe musical, *Brigadoon* reached the screen in 1954 starring Gene Kelly. But movie audiences didn't have an opportunity to see *Finian's Rainbow* until 1968—twenty-one years after it first appeared on Broadway. The long delay was not caused by a lack of trying.

Finian's Rainbow had a score by E. Y. Harburg and Burton Lane, with book by Harburg and Fred Saidy. It opened in New York at the 46th Street Theater in January of 1947, under the direction of Bretaigne Windust. Choreography was by Michael Kidd.

The delightful, socially-aware fable deals with an Irishman named Finian (Albert Sharpe) and his daughter, Sharon (Ella Logan) who journey to the American South—specifically to Rainbow Valley in the state of Missitucky. Finian is followed to the United States by Og (David Wayne), a leprechaun. Og must retrieve a crock of gold the Irishman stole from him and his fellow sprites. The crock has magic powers to fulfill three wishes.

Sharon falls in love with Woody Mahoney (Donald Richards), a local sharecropper, and Og becomes smitten with Susan (Anita Alvarez), Woody's deaf-mute sister, who communicates through her dancing.

In a moment of anger, Sharon uses one of the crock's wishes to turn the play's villain—a bigoted senator—into a black man. Later, he is changed back to his natural pigment, but he has now become more liberal. The third wish gives Susan the powers of hearing and speech.

The story ends with Woody perfecting a home grown mentholated tobacco, thus insuring Rainbow Valley's prosperity; Og becoming human; and the wish-drained crock losing its value. Finian decides to move on, leaving his daughter to marry Woody.

The score of *Finian's Rainbow* contains many popular standards. Among the Harburg/Lane numbers that achieved greatness outside of the show were "How Are Things in Glocca Morra?" "Old Devil Moon," "Look to the Rainbow," "If This Isn't Love," and "When I'm Not Near the Girl I Love," which was engagingly delivered by David Wayne. Other songs, like "When the Idle Poor Become the Idle Rich," "The Begat," and "Necessity," carried a message. Albert Sharpe was properly

Finian's Rainbow. Petula Clark and Fred Astaire.

Finian's Rainbow. **Petula Clark, Don Francks, and Fred Astaire.**

roguish as Finian, while Miss Logan exhibited an abundance of bounce and enthusiasm as Sharon. Her rendition of "Glocca Morra" was one of the most stirring moments of the evening.

John Chapman in the New York *Daily News* called the musical "captivating whimsey . . . . Mr. Harburg has written the lyrics and they are top grade —witty or pretty, funny or satirical as needs dictate. Burton Lane has set them to pleasant melodies. And, under the guidance of Michael Kidd, an inspired white and Negro chorus does some of the best dancing imaginable."

The original production of *Finian's Rainbow* ran for 725 performances and received the Donaldson award for best musical of the season. There were numerous revivals over the years. The play has, in fact, been a favorite in tent, dinner, and community theaters across the nation.

In 1967 Joseph Landon, who was then producing the Warner Brothers film adaptation of *Finian's Rainbow,* discussed with the *Los Angeles Times* the reasons it had taken the musical so long to be made into a movie. "The lyrics were by Yip Harburg," he said, "and this was his love child. It was Yip who set the price tag on the play; he wanted one million dollars for the film rights and total control of the material.

"But by this time [in the late forties] he was into the McCarthyism period and nobody would touch *Finian* with a 10-foot pole. Not so much because of the supposedly radical nature of the lyrics ["When the Idle Poor Become the Idle Rich"] as because of the civil rights aspects, the prominence of southern sharecroppers and their integration in the story."

About seven years after its Broadway debut, it looked as if *Finian* might finally get off the ground as

Finian's Rainbow. Fred Astaire.

Finian's Rainbow. Tommy Steele and Keenan Wynn.

Finian's Rainbow. Don Francks and Petula Clark.

Finian's Rainbow. **Barbara** Hancock and Fred Astaire.

a film—an animated one. Distributors Corporation of America acquired the rights and recorded the musical track to what was intended to be a feature length cartoon. Even though the singing voices included the likes of Frank Sinatra, Ella Fitzgerald, Louis Armstrong, and Judy Garland, the project was finally abandoned—after one million dollars had been spent.

There were a number of other scuttled attempts over the years. Mickey Rooney wanted to do Og when he was under contract at Metro; Debbie Reynolds was announced for a 1960 film to be directed by Michael Gordon; and both Dick Van Dyke and Burt Lancaster were mentioned for other possible productions.

Finally, in 1966, Landon brought the idea of making the picture to Warner Brothers executive Ben Kalmenson. Kalmenson was interested—particularly if Fred Astaire could be signed to star.

Recalled Landon: "As a starter, we luckily both had the same agent. I send Fred the album and the libretto and he loved them—'but as Finian,' he asked, 'what do I do?' It was a good question—I myself had long felt that the play was flawed dramatically in one way: Finian, the leading character, is a non-musical character; he neither sings nor dances."

With the blessings of Harburg and Saidy, who wrote the screenplay, and director Francis Ford Coppola, producer Landon set out to remedy the problem. Adjustments in the script were made so that Astaire took part in both the "Look to the Rainbow" and "Idle Poor" numbers.

"This is the first time," remarked the producer, "that Fred isn't really playing Fred. He's Finian, arthritis and all—and he's even worked arthritis into his dancing."

Director Coppola, who was still several years

away from his history-making *The Godfather*, rehearsed his cast intensely for six weeks, then had them appear in a one-night theatre-in-the-round performance as a tryout before commencing with principal photography. As a result he was able to bring the movie in for considerably under its announced four million dollar budget.

Joining Astaire in the cast were Petula Clark as Sharon, Tommy Steele as Og, Don Francks as Woody, Keenan Wynn as the senator, and Barbara Hancock as Susan the Silent. In a part considerably expanded from the original, Al Freeman, Jr., had the film's funniest moment as an educated black botanist, who tongue-in-cheeks the cliché Stepin Fetchit shuffle for Wynn's irritated benefit.

All the songs from the original play—save "Necessity"—were retained for the picture version.

The 1968 release had much to recommend it, but there were also some serious flaws. On the plus side was the always delightful and seemingly effortless performance of Fred Astaire, dancing down the road like a Gaelic Pied Piper. Miss Clark, too, was appealing and delivered fine renditions of all her numbers.

Visually, *Finian's Rainbow* had some grand moments, particularly in a beautiful under the credits sequence in which Astaire and Miss Clark—with a chorus singing "Look to the Rainbow"—journey on foot across scenic America. "The Begat," in which a now-black Keenan Wynn and three negro singers travel cross-country in a car, and the cavorting "If This Isn't Love" were two other numbers in which Coppola made full use of the film medium, improving on the staging by having his camera and editing complement the choreography of Hermes Pan. Singing and dancing in the production was quite good and energetic.

Perhaps the biggest problem in the movie was the performance of Tommy Steele as Og. A gifted entertainer, Steele, in this instance, was allowed to overplay the leprechaun to the point of being annoying. It was a portrayal completely out of balance with the rest of the cast.

Another irritant was the poor mixture of exterior and interior settings. At its best when out-of-doors, the film was jarring when, during its night sequences, it moved onto a nicely designed but very unreal sound stage exterior.

When *Finian's Rainbow* first appeared back in 1947, its civil rights aspects were, to say the least, controversial, however most critics in 1968 found them quaint—even though some of the attitudes had been updated. In general, most reviewers panned the picture, naming Coppola's emphasis on camera movement as a principal cause of the failure.

"Filming *Finian's Rainbow* in wide screen diminishes the musical rather than enlarging it," said *Boston After Dark*, "and the gaiety and the underlying sadness that tugged at the corner of one's mind simply disappears under the onslaught of the movie's stupendous technology."

The Warner release was a box office bomb. With the exception of a brief co-host appearance with Gene Kelly in *That's Entertainment, Part II*, it was Fred Astaire's last role in a screen musical.

19
Oliver!
(1968)

The work of Charles Dickens has served as the basis for several musical stage and/or screen productions. *The Pickwick Papers, The Old Curiosity Shop, A Tale of Two Cities,* and A *Christmas Carol* have each been peppered with songs and dramatized in one medium or the other. Certainly the most successful musicalization of the author's work was *Oliver!,* which had its debut on London's West End on June 30, 1960.

Based on *Oliver Twist,* this hit musical (2,618 performances at London's New Theatre) by Cockney writer / composer Lionel Bart starred Ron Moody as Fagin and Georgia Brown as Nancy. It was imported to the United States in 1962 by producer David Merrick, and made its American debut in Los Angeles, then went on to San Francisco, Toronto, and Detroit before its initial Broadway appearance at the Imperial Theater on January 6, 1963. Peter Coe was director of the production. Miss Brown journeyed to the U.S. to reprise her original role, but Clive Revill replaced Moody as Fagin. Also in the New York company were Bruce Prochnik (Oliver), David Jones (The Artful Dodger), and Danny Sewell (Bill Sikes).

Oliver! is a free adaptation of the Dickens novel. It picks up young Oliver Twist when he is an inmate in the sinister workhouse run by Mr. Bumble, follows him through his very brief apprenticeship with the undertaker, Mr. Sowerberry, and on to his adventures in London where he becomes a member of Fagin's school for young thieves. There he meets Fagin's prize pupil, The Artful Dodger, a hardened criminal named Bill Sikes, and Bill's warm and friendly girl friend, Nancy.

Master Twist was never cut out for a life of crime. Arrested by mistake before he even has a chance to attempt any nefarious activities, he is later adopted by a wealthy gentleman and given a good home. Sikes, Nancy, and Fagin, fearing Oliver will reveal what he knows about their illegal activities, kidnap the lad. Nancy, however, in a change of heart, tries to return Oliver to his benefactor, but is killed by Sikes. He, in turn, dies while trying to escape. Fagin, having lost everything, decides to join with the Dodger and start anew. Oliver rejoins his benefactor and presumably lives happily ever after.

Staged with an extreme of stylization, *Oliver!* had little choreography and utilized its mobile sets (designed by Sean Kenny) to transport characters from one scene to the next. Composer Bart supplied a superb cache of songs—all pleasant to the ear— that set mood, defined character, or just plain

Oliver!. Mark Lester is about to ask for more "Food, Glorious Food."

inspired audiences to tap their feet in time with the music. "Food, Glorious Food" is the workhouse orphans' lament, while Oliver expresses his seemingly hopeless plight in the moving, "Where Is Love?" The Dodger welcomes Oliver to London with "Consider Yourself," and a few minutes later Fagin explains the facts of life to him in the delightful "You've Got to Pick a Pocket or Two." Fagin has another fine song with his introspective "Reviewing the Situation."

Nancy also has some choice numbers. "As Long As He Needs Me," in which she expresses her feelings toward Sikes, became a major record hit. "I'd Do Anything," performed along with Fagin's boys, spoofed polite society, and "It's a Fine Life" and "Oom-Pah-Pah" were two lusty tunes she sang in the music hall style.

John McClain of the New York *Journal American* called the play "simply scrumptious . . . Never before have so many gifted urchins filled a stage to better effect."

Oliver! played a total of 774 performances.

It took seven years from its London debut for Lionel Bart's musical smash to reach the screen. Director Richard Quine was among the first to express an interest in the film rights. He hoped to do the picture in conjunction with the Mirisch Corporation, and star Peter O'Toole as Fagin and Miss Brown as Nancy.

About a year later, in 1964, it looked as if Peter Sellers was going to play Fagin in a co-production between his company and Columbia Pictures.

Three months passed and producer John Woolf announced that he had the rights to the property. He planned to approach the Richard Burtons (Elizabeth Taylor) to portray Nancy and Sikes. If all went well, Laurence Harvey would be Fagin. Obviously that

Oliver!. Jack Wild, Ron Moody, and Mark Lester.

casting plan didn't materialize either, because, in December of that year—within a ten day period—both Peter O'Toole and Sean Connery were mentioned by Hollywood columnist Mike Connolly as being set to play Bill Sikes. In addition, at one point Bryan Forbes was announced as the movie's probable director.

When the production finally stabilized, Woolf, through his Romulus Productions, had the property and planned to release it through Columbia. Music supervisor John Green, designer John Box, and choreographer Onna White were signed to the project from the start. Vernon Harris did the screenplay, and Sir Carol Reed, who had never before done a musical, was signed as director.

"Reed wanted to go back to the essence of Dickens," recalls John Green. "He began with the original novel, then worked up through the Bart concept. On stage, *Oliver!* was done in a very stylized manner, but the movie opened up the show considerably and played it on a realistic level. At one point, Bart thought we were destroying his play.

"We also had a fundamental difference in the *Oliver!* music. In the theater, a fourteen-piece orchestra was used to give a chamber opera effect but, on film, we had a seventy-four piece symphony orchestra."

Nightclub singer Shani Wallis, who beat out Georgia Brown as Nancy, was the first person cast in the movie. "I was living in New York," she says, "and went back and forth to Hollywood for various tests for almost a year before I knew I had the part.

"The producers kept experimenting. They dyed my red hair black—because Georgia's was black, I guess—then had me go back to the original red."

As Fagin, producer Woolf and director Reed went back to the stage original and signed Ron Moody, who reflected in the *Los Angeles Times:* "I

Oliver!. Shani Wallis (r) sings "It's a Fine Life."

played it very Jewish on the stage, but we changed it for the film. My stage Fagin caused no uproar at all, but I didn't think he was right for the film and Sir Carol Reed agreed. He's not terribly kosher now. Attitudes have changed. I play him kind of mockingly because I think it's healthy for us to realize that what was once anti-Semitic is now best handled by a light approach."

Oliver Reed was cast as the menacing Bill Sikes, Jack Wild was The Artful Dodger, Harry Secombe played Mr. Bumble, and Mark Lester got the play's title role. Budgeted at approximately ten million dollars, the picture shot in England at the Shepperton Studios.

"Reed was a marvelous director and a very stubborn man," reports Miss Wallis. "Even though he'd never directed a musical before, he followed his instincts—which were usually right—rather than blindly following the well-meaning suggestions of

Oliver!. **Oliver Reed and Ron Moody.**

Oliver!. The "Who Will Buy?" number.

those who were more experienced with musical productions.

"He turned many musical situations around and took songs that were just 'sung' on stage and used them in a more dramatic sense. For example, in my 'Oom-Pah-Pah' number, he had me steal Oliver away from Sikes while I sang and danced. That sort of thing wasn't done on stage."

All but two tunes from the stage production were retained for the movie. One of the missing numbers, "My Name," was reportedly dropped because Oliver Reed, whose song it was, was not a singer.

Few films have captured the true look and atmosphere of Dickens' London as did *Oliver!* Designed by John Box, the realistically detailed streets and interiors in muted colors were stunning, as were Phyllis Dalton's splendid costumes. And Onna White's sprightly dance ensembles—particularly in the "Consider Yourself" and "Who Will Buy?" sequences—utilized choruses of hawking news boys, buskers, peddlers, and chimney sweeps, giving a panoramic view of London street life.

Moody was the perfect Fagin, making an effective entrance through a smoke-filled doorway, and scurrying about the back alleys like a river rat. Miss Wallis, an appealing lady with a fine singing voice, endowed Nancy with just the right amount of bawdiness, and Reed was properly brutish as Sikes. Mark Lester evoked the necessary sympathy as Oliver, but the real scene-stealer of the film was Jack Wild, . who performed like a veteran of the music halls and proved himself to be one of the most talented child stars to come along in some time.

Carol Reed, one of the screen's finest dramatic directors, emphasized characterization in his film, and as a result made the Dickens people come to life much more than they had on the stage.

"It's a bright, shining heartwarming musical," said *Variety,* "packed with songs and lively production highspots."

Oliver! was the big winner at the 1968 Oscar ceremonies. The picture, which would earn a domestic rental gross of $16.8 million, won the awards for best picture, director, art direction, scoring (of a musical), sound, and a special statuette to Onna White for her brilliant choreography.

Oliver! is as near perfect as a musical can get. It is a careful blending of realistic and stylized elements into a film that entraps its audience from start to finish. An excellent family picture, it easily ranks alongside *Mary Poppins, The Wizard of Oz,* and other timeless classics of the genre.

Oliver!. **Jack Wild and Mark Lester.**

20
Hello, Dolly!
(1969)

Like his *On the Town*, Gene Kelly's movie version of *Hello, Dolly!*, the second most successful musical in Broadway's history, was a landmark production—of sorts. It has the dubious distinction of being a film that nearly broke Twentieth Century-Fox.

Hello, Dolly! was a piece of musical theatre devised for a single purpose—to entertain. There were no profound messages, no tragic scenes to bring forth tears. All that producer David Merrick intended was that the audience leave the theater feeling good, and, possibly, humming the title song.

The show made its debut in January of 1964 at the St. James Theater. Its book was by Michael Stewart, adapted from Thornton Wilder's play, *The Matchmaker*, produced a decade earlier starring Ruth Gordon and subsequently made into a movie with Shirley Booth. Jerry Herman wrote the nostalgic music and lyrics for *Dolly*, and Gower Champion directed.

When Ethel Merman proved unavailable to play the leading role in this old-fashioned, "Gay Nineties" musical, banjo-eyed Carol Channing assumed the part of the widowed New York match maker, Dolly Levi, endowing it with her own special brand of innocence, boisterousness, and charm. There's no doubt that, despite the many fine actresses who have played Dolly since, that role will always belong to Miss Channing.

Story-wise, the show deals with marriage broker Dolly's clandestine plan to snag rich blustering Yonkers storekeeper Horace Vandergelder (David Burns) as her second husband. However skinflint Vandergelder is interested in New York milliner Irene Malloy (Eileen Brennan). Dolly inspires the businessman's two naive store clerks, Cornelius (Charles Nelson Reilly) and Barnaby (Jerry Dodge), to close the store and call on the attractive Miss Malloy and her cute assistant, Minnie (Sondra Lee), knowing that they will be discovered there by Vandergelder. Her maneuverings are not intended to be malicious but to create confrontations which she, in her infinite wisdom, knows will—in the long run—make everyone happy. The inexperienced young men will gain some independence and be matched with two ladies who will be right for them, and she will get the opportunity to be a good wife for Horace.

The climactic scene in this musical farce occurs in the posh Harmonia Gardens restaurant. Once Horace discovers his clerks there and learns that they are on a one-day adventure in New York, he

Hello, Dolly. **Carol Channing in the original Broadway production.**

explodes, and it's total pandemonium. Vandergelder winds up in jail. Upon his release, he mends his selfish ways. He makes Cornelius a partner and asks Dolly to marry him.

Aside from the foot-tapping title tune, which is certainly one of the greatest show-stopping numbers ever to come out of the musical theater, there were several other enjoyable moments of song in *Hello, Dolly!* Vandergelder gave audiences a laugh when he revealed his motivations for wanting to get married in "It Takes a Woman," while Cornelius and Irene shared the tender love song, "It Only Takes a Moment." "Put on Your Sunday Clothes," "Dancing," and "Elegance" were the energetic numbers that featured various combinations of cast members, and Miss Channing had two good songs all to herself—"So Long, Dearie" and "Before the Parade Passes By."

Splendid costumes and settings, Champion's wild, exciting choreography, an endearing score comprised of waltzes, polkas, and quartet pieces, and the performances of a talented cast combined to make what was essentially a thin property into a warm and wonderful evening's entertainment.

"It transmutes the broadly stylized mood of a mettlesome farce into the gusto and colors of the musical stage," said Howard Taubman in the *New York Times.* "What was larger and droller than life has been puffed up and gaily tinted without being blown apart. *Hello, Dolly!* is the best musical of the season thus far."

The show won the Tony and New York Drama Critics Circle awards as best musical of the season. It played a total of 2,844 performances—with several different casts—before it finally closed. It was this lengthy run that contributed to the financial woes of the motion picture version.

In 1965, Twentieth Century-Fox purchased the screen rights to *Hello, Dolly!* for two million dollars. This price in itself was no record, considering what Jack Warner had paid for *My Fair Lady.* What made the *Dolly* deal unique was that the studio also guaranteed David Merrick that it would not release its movie until the original Broadway production closed, or until June 20, 1971—whichever came first.

Produced and written for the screen by Ernest

Lehman, the nineteen million dollar plus film was completed in mid-1968, but, alas, Merrick's stage production was still going strong. The picture sat on the shelf for about a year while the studio paid interest on its financing loan at the approximate rate of one hundred thousand dollars per month. In order to get its product out into the marketplace and earning money, Fox finally agreed to pay Merrick an additional one million dollars in exchange for the right to release the movie at the end of 1969. It's estimated that by the time of the gala premiere, the studio had close to twenty-five million invested in *Hello, Dolly!*.

Even before hiring a director for this mammoth project, Fox had set about to sign a leading lady. As with all movies based on well-known literary properties, there was much speculation around Hollywood as to who would play Dolly Levi. Carol Channing was, of course, the sentimental favorite, while some insiders were predicting that the plum would go to Elizabeth Taylor. Then the studio fooled everybody. It announced that Barbra Streisand would star in its production of *Hello, Dolly!*

"With all due respect to young Miss Streisand," wrote Richard L. Coe in the *Washington Post*, "the mournful Nefertiti is clearly not the outgoing zestful Irish woman whose vitality brightens Thornton Wilder's mature, life-loving Dolly Gallagher Levi.

"The perversity of not choosing to get Carol Channing's musical comedy classic on film is hard to fathom. While others have played Dolly on stage, hers is the only one which, in four years, never has played to an empty seat."

The producer's reasoning in not choosing Miss Channing was at least easy to understand—if not wholly to accept. As had been the situation of Julie Andrews in *My Fair Lady* and Ethel Merman in *Gypsy*, Carol Channing had no track record in motion pictures. But then, since Miss Streisand's *Funny Girl*, which she'd recently done at Columbia, had not yet been released, she didn't have a proven track record either.

"All top executives here concurred with me that Barbra Streisand is the best choice we could make for the film version of *Dolly*," said producer/writer Lehman in defense of the widely-criticized casting.

Hello, Dolly. **Walter Matthau and Barbra Streisand.**

Hello, Dolly. Walter Matthau, Michael Crawford, and Danny Locklin sing "It Takes a Woman."

Hello, Dolly. Walter Matthau and Barbra Streisand.

Refering to the performer's stage and television appearances, he added, "She comes over not larger than life, but real.

"I'm rather amused, in fact, by the manipilated anti-Streisand campaign represented in letters received here at Twentieth, because of their startling similarity in wording. It's just too much of a coincidence, even though they come from various parts of the country."

Miss Streisand was considerably younger than any of the other actresses who'd played Dolly, however Lehman pointed out to the *Los Angeles Times:* "Nothing in my screenplay will indicate that Dolly had been married for X years and widowed for X years. As a matter of fact, in Thornton Wilder's original play, *The Matchmaker,* he noted all character's ages except that of Dolly, which he set down as 'uncertain age.' She could well be a widow in her thirties."

Lehman's choices for a director, choreographer, and co-star for Streisand garnered far less negative comment than had her casting. Indeed, they were widely approved. Gene Kelly was entrusted to direct the film and Michael Kidd was signed as choreographer. As Horace Vandergelder, the producer set Walter Matthau-a perfect selection. Other castings included British actor Michael Crawford as Cornelius, Marianne McAndrew as Irene Malloy, E.J. Peaker as Minnie, and Danny Locklin as Barnaby Tucker. Louis "Satchmo" Armstrong, who'd made a runaway hit recording of "Hello, Dolly!" was brought in to sing a chorus with Barbra in that big production number on the Harmonia Gardens set.

Gene Kelly's *Hello, Dolly!* was a big, expansive, pictorially opulent production—the sort of film one would expect from such a distinguished graduate of Metro-Goldwyn-Mayer's musical "university." It featured a two million dollar recreation of early New York City, a parade of thousands of gaily costumed extras, and entailed the conversion of upstate Garrison, New York, to represent Yonkers of the 1890s. Jerry Herman wrote two new songs for the show ("Just Leave Everything to Me" and "Love is Only Love"), which were no better nor worse than some tunes eliminated from the original score ("I Put My Hand In," "Motherhood").

The picture failed on several counts. Great talents as they might be, Kelly and Kidd could not spark their movie with the excitement that had been

Hello, Dolly. **Waiters sing the show-stopping title tune to Barbra Streisand.**

Hello, Dolly. Louis Armstrong and Barbra Streisand.

Hello, Dolly. Walter Matthau and Barbra Streisand.

Hello, Dolly. Walter Matthau and Barbra Streisand.

generated on the stage. The dances were lively, often exaggerated, but they nevertheless seemed overlong and flat with little variance of perspective.

Gene Kelly in a 1968 interview with the *Los Angeles Times:* "*Dolly* couldn't be done the way Gower Champion did it on the stage—this light piece, this commedia dell'arte—because the camera couldn't accommodate it. We decided to play it for real. . . . Actually, it's an old-fashioned picture, based on the manners and the mannerisms of the people, on the performances of the actors. If Walter and Barbra weren't giving believable performances, we'd be lost. The idea is to make the picture come alive through the performances. It's a movie-movie. I wish I could say we're breaking new ground, but we're not."

What Kelly did wind up with on the screen was, unfortunately, a mixture of comedic styles—a combination of French farce and Mack Sennett slapstick—that didn't really work. He kept his action moving along at a nice pace, yet never seemed able to get his audience involved in what was going on.

Perhaps the best sequence in the picture was the pre-title opening on a still of the imaginatively detailed New York street, which presently wiped into live action. The camera then momentarily concentrated on the tapping, dancing feet of passersby, ultimately picking up Dolly on her way to the railway station and Yonkers. This was the kind of moment fans of Kelly and Kidd would expect, but there was little cleverness of this sort in the rest of the movie.

The casting of Barbra Streisand was a mistake. She did her cute shtick, sang in her marvelous individual style, and apparently tried very hard. But she was *not* Dolly. The general consensus was that she was doing a poor imitation of Mae West. Walter Matthau, on the other hand, was nothing short of brilliant as Vandergelder, stealing every scene he was in.

In his *New York Times* review, Vincent Canby said, "Gene Kelly, who directed two classic musicals with Stanley Donen *(Singin' in the Rain* and *On the Town),* here acts like a caretaker of a big, valuable property. He and Michael Kidd, his choreographer, have protected everything Gower Champion gave the original, and added nothing to the heritage of the musical screen except statistics."

Somehow, *Hello, Dolly!* managed to snag a best picture nomination come Academy Award time. It lost to *Midnight Cowboy,* but did win statuettes for art direction-set decoration, sound, and scoring of a musical. Its domestic box office gross—$15.2 million—was disappointing to the powers at Fox, especially since some of their other recent big budget efforts—*Dr. Dolittle, Star!* and *Tora! Tora! Tora!*—had proved to be financial bombs also.

One additional reason for *Dolly's* box office failure might be the fact that it took so long (five years) for it to reach the screen. A show like *Oklahoma!* certainly took much longer, but by the time it was immortalized on celluloid it was considered a classic and the Rodgers and Hammerstein songs were part of our musical heritage. Conversely, with virtually every major recording artist having made a disc of "Hello, Dolly!", the public was deluged with that number over the five year period and it's quite possible that, by the time the movie opened, people were, frankly, bored with the show.

21
Paint Your Wagon
(1969)

"The play's the thing," as Shakespeare said, and no vast amount of money spent on lavish production values is going to rescue a film strapped with a muddled screenplay.

Even in its stage version, *Paint Your Wagon* had script problems. It was the second collaboration of Alan Jay Lerner (book and lyrics) and Frederick Loewe (music), who in 1947 had contributed *Brigadoon* to the musical theater. Cheryl Crawford produced the work, which opened in November of 1947 at the Shubert Theater. Daniel Mann directed and Agnes deMille choreographed a cast headed by James Barton, Olga San Juan, and Tony Bavaar.

Rich in detail of western Americana, *Paint Your Wagon* was set in 1853, during the California Gold Rush. Ben Rumson (Barton) is a prospector who lives with his daughter Jennifer (Miss San Juan) in the mining camp he founded, Rumson Creek. When a gold nugget is discovered, prospectors move to the town in droves.

Jennifer becomes enamoured with a young man of Mexican descent, Julio (Bavaar). To protect the girl from him and the other miners in this town lacking in women, Ben sends her east for schooling. Then he buys a Mormon woman for himself at an auction.

When the gold is played out at Rumson Creek, the once-booming camp is left a ghost town. Jennifer returns home shortly before her father's death. She takes up again with Julio, and vows that Ben's dream for his town will not die. Rumson Creek will survive as a farming town.

The strongest asset in *Paint Your Wagon* was the score. There were western-like folk songs ("Wanderin' Star," and the hit "They Call the Wind Maria"), a couple of knee-slapping dance sequences ("Hand Me Down That Can o' Beans" and "Whoop-Ti Ay"), the chorus-building "I'm on My Way," and two fine sentimental ballads, "I Talk to the Trees" and "I Still See Elisa."

The book was another matter entirely. Containing little humor and a weak second act, Lerner's script dampened what should have been a major stage success. Commenting in the New York *Herald-Tribune,* Walter Kerr said, "Writing an integrated musical comedy—where people are believable and the songs are logically introduced—is no excuse for not being funny from time to time. But the librettist of *Paint Your Wagon* seems to be more interested in the authenticity of his background than in the joy of his audience."

The show closed after 289 performances.

Louis B. Mayer, ousted MGM production chief, bought the film rights to *Paint Your Wagon* early in 1952. Paying two hundred thousand dollars plus a percentage, Mayer outbid both Paramount and the studio he'd headed for twenty-seven years. The picture never went before the cameras, however, and following the mogul's death in 1957, *Wagon* was tied up in his estate.

In 1964, singer Eddie Fisher optioned the rights for fifty thousand dollars and announced that he would film it independently in Cinerama. He would play Julio himself and hoped that James Cagney would be lured out of retirement for the James Barton role. A year passed. No firm production plans were set. Cagney had obviously turned the project down, because *Variety* ran a story stating that Mickey Rooney was after the Ben Rumson part.

Fisher, unable to get his picture going, finally let his option drop. The property was picked up by Paramount and definite production plans put into motion: Alan Jay Lerner was to produce the film, Joshua Logan was to direct, and Lerner was to do the screenplay, from an adaption by Paddy Chayefsky of the original Lerner/Loewe stage book. Additional songs by Lerner and Andre Previn were also to be included.

Since the problem with the stage play was the script, it's not surprising that Lerner and Chayefsky chose to abandon virtually the entire original story and start fresh. They came up with a sexy morality play about Ben Rumson, a drunken, brawling prospector who strikes gold while digging a grave to bury a dead pioneer. Gone are the daughter and her Spanish lover. Instead, Rumson has a partner named "Pardner," whose job it was to see that the older man's debts are paid and carry him home whenever he should happen to fall drunk in the muddy streets of the mining camp—No-Name City.

The principal female in the newly-fashioned plot line is Elizabeth, the second wife of a traveling Mormon, whom Rumson buys at auction. He falls in love with her and allows her to reform him.

Paint Your Wagon. **Lee Marvin and Clint Eastwood.**

Paint Your Wagon. Clint Eastwood and Jean Seberg.

Paint Your Wagon. Lee Marvin.

Knowing that his fellow prospectors covet his wife, Rumson kidnaps a stagecoach full of French prostitutes, bringing them back to No-Name City. In his absence, however, Elizabeth and Pardner have also become smitten with each other. The partners agree to do the unorthodox thing—live together and share Elizabeth.

Later, the heroes hit on a plan to tunnel under the town's saloons to retrieve the loose gold dust that falls through the floorboards—thus eventually undermining the town. When No-Name City collapses into the tunnels, everyone, including Rumson, moves on. Pardner decides to remain with Elizabeth and become a farmer.

Casting of *Paint Your Wagon* raised a few eye brows in Hollywood circles—mainly because almost none of the performers were singers. Lee Marvin was paid one million dollars plus a percentage to play Rumson. He maintained it would be his first and last singing role: "You can play Walter Huston only once." Clint Eastwood was Pardner, Jean Seberg did Elizabeth, and Harve Presnell—the only good voice in the bunch—was signed on to deliver "They Call the Wind Maria."

All the best numbers from the Lerner/Loewe score were retained, proving to be the strongest musical material in the picture. Previn's songs were functional but forgettable. Commenting to the *Los Angeles Times* about the tone of his lyrics, Lerner said, "After fourteen years, going back over it, I suddenly realized that all the songs dealt with loneliness. But in the original show, 'I talk to the Trees' was a happy ballad. Here I've rewritten some lyrics and it's a comment on Clint Eastwood's lonely condition."

The picture began filming in summer of 1968 outside the difficult-to-get-to town of Baker, Oregon, doubling for northern California. It was a rough shoot, with some poor weather and difficult logistics slowing down the production timetable. Rumors began filtering back to Hollywood that director Joshua Logan was in trouble. Scenes were having to be re-shot, and he was not a strong enough director to handle Lee Marvin, who, the tales continued, had not been entirely "on the wagon." The fact that Logan's last film—*Camelot*—had received a less than enthusiastic reception from cirtics didn't help his case any. Talk was that Richard Brooks, who'd directed Marvin in *The Professionals,* might fly up to Oregon and replace Logan.

Paint Your Wagon. **Clint Eastwood.**

Logan stayed with the seventeen million dollar picture and, later, in an interview with *Variety,* discussed the rumors: "Everyone was speculating about how to do it better, but no official moves were made to me about leaving the project; no one ever said a word to me. But, everyone goes into a state of shock over an ambitious film. This one was no different from the average big scary picture.

"There has always been basic agreement on what we are trying to do, but since boyhood, I've always had arguments with actors and authors over proper interpretation. This is normal in any theatrical environment."

Running 166 minutes plus intermission, *Paint Your Wagon* debuted late in 1969. Reviews were not bad, but considering the money and talent that went into this production, they weren't really good either. Vincent Canby in the *New York Times* said; "There is something quite cheerful about its book, which is so casual that it stops being a story after intermission and becomes, instead, a frame for some amusing comedy 'set pieces.'" Lerner, even with the help of Chayefsky, was still not able to create a compelling story from which to make a solid film.

As delivered by the three principals, the Lerner/Loewe score was disappointing. Good music needs good voices, not merely good actors with audience appeal. A bit of realism should have been sacrificed here in favor of dubbed voices. Almost all the film's musical numbers, in fact, seemed to simply just happen, and were not particularly well integrated into the story. There were no standout dance numbers.

The movie's pluses included the big, beautiful Oregon scenery, a detailed recreation of the Gold Rush period complete with hundreds of extras and a marvelously eccentric performance by Marvin, reminiscent of the part he played in *Cat Ballou*. His comedy bits may have been slightly broad at times, but they were certainly funny.

Paint Your Wagon was not a box office success, grossing only $14.5 million in domestic rentals. It was one of those misconceived productions that counted too much on spectacular production values and star power to replace key factors like properly executed musical numbers and a good, well constructed story. Quite an expensive lesson in filmmaking.

Paint Your Wagon. **Harve Presnell sings "They Call the Wind Maria."**

22
Cabaret
(1972)

"Today I get very antsy watching musicals in which people are singing as they walk down the street or hang out the laundry," director Bob Fosse told *New York* magazine in 1974. "In fact, I think it looks a little silly. You can do it on the stage. The theatre has its own personality—it conveys a re moved reality. The movies bring that reality closer."

Fosse's 1972 film of *Cabaret* is, to date, the most convincing argument for the unintegrated musical, one in which songs and dances are separated from the dramatic action. All of its songs—with the exception of one in a beer garden—are performed on the cabaret stage.

Reminiscent at times of Kurt Weill and his *The Three-Penny Opera*, *Cabaret* was adapted from John van Druten's award-winning play, *I Am a Camera* (1951), which in turn was based on Christopher Isherwood's *Berlin Stories*. Produced by Harold Prince with Ruth Mitchell and directed by Prince, the musical opened in November 1966 at the Broadhurst Theater. Joe Masteroff did the book, Fred Ebb wrote the lyrics, and John Kander composed the music. Choreography was by Ronald Field.

The play tells of an American writer/teacher, Clifford Bradshaw (Bert Convy), who comes to Berlin in 1929. He finds a city full of moral decay and frightened lost souls, rushing blindly toward its Armageddon—Nazism.

The writer also encounters a series of interesting characters—each a sad product of the time and conditions in which they exist. The landlady of his rooming house, for example, is Fraulein Schneider (Lotte Lenya). She plans to marry another of her tenants, Herr Schultz (Jack Gilford)—a Jew—until she realizes that such a union might be politically unwise. Ernst Ludwig (Edward Winter), on the other hand, is a Nazi, whom Cliff at first befriends and then battles both ideologically and physically. Then there is Sally Bowles (Jill Haworth), a mediocre singer at the decadent Kit Kat Club. She's a vulnerable, truly naive, girl who has known many men and, although she says she loves Cliff, is obviously bent on destroying both the relationship and herself.

The most fascinating of the play's characters was the nightclub Emcee, brilliantly delineated by Joel Grey. His was a symbolic character, representing all that was vulgar and ugly in Germany of the early thirties. It is through this role that the audience feels the frightening advance of the Nazi juggernaut.

Cabaret, like *I Am a Camera*, concludes with

Cabaret. **Joel Grey and company.**

Cliff leaving Sally behind in the doomed Germany. She has decided to continue her wild lifestyle, oblivious of the ominous fate that awaits her.

"Willkommen" was the song with which the Emcee greeted the theater audience each night, telling them to relax and have fun, because here in the cabaret life is without care. Then there were such amusing and/or poignant numbers as "Meeskite," Herr Schultz's intoxicated ramblings about his ugly face. "If You Could See Her," "The Money Song," "Two Ladies," and "Don't Tell Mama" were tunes sung in the nightclub by the Emcee, Sally, or both—all conveying either character insight or a reflection of the period.

The title tune, "Cabaret," was the show-stopper that expressed Sally's irresponsible philosophy of life. It went on to become a major hit. Yet possibly a more interesting musical spot in the show occurred at the end of the first act when the Kit Kat waiters stepped forward to sing the initially pastoral, then subtly threatening, "Tomorrow Belongs to Me," a hymn to the Fatherland.

Cabaret was an engrossing, disturbing theatrical experience. Harold Prince kept his staging fluid, and Ron Field's choreography, especially in the energetic "Telephone Song," was continuously vibrant. Flashy atmospheric settings by Boris Aronson and Patricia Zipprodt's ribald costumes were among the most imaginative seen on Broadway in many seasons. Performances in general were highly praised, particularly those of Lotte Lenya, Jack Gilford, Bert Convy, and Joel Grey. The production was awarded the Tony and New York Drama Critics awards for best musical, and played 1,166 performances before it finally closed.

"*Cabaret* is a stunning musical," said Walter Kerr in the *New York Times*. "This marionette's-eye view of a time and place in our lives that was brassy, wanton, carefree and doomed to crumble is brilliantly conceived."

The original deal for movie rights to *Cabaret* was with Cinerama, Inc., for a reported price of over $2.1 million dollars. But that situation fell through

Cabaret. Liza Minnelli and Marisa Berenson.

and the property was ultimately acquired by ABC Pictures for a co-production deal with Allied Artists. The price on that deal was only $1.5 million plus a participation once the profits reached a certain figure.

Cy Feuer produced the 1972 Technicolor release, which had a screenplay by Jay Allen and Hugh Wheeler. The director was Bob Fosse.

"*Cabaret* is vastly changed from the Broadway show," Fosse told *After Dark* following the film's release. "It's only my opinion, but I think we've retained the better parts of the show and dropped the weaker. I always felt the stage book was very weak, but the cabaret numbers and that cabaret atmosphere were incredible. And the overall concept was marvelous."

As previously noted, Fosse eliminated all the numbers save one that did not take place in the cabaret. The lone survivor was "Tomorrow Belongs to Me," which provided a chilling moment in the picture when it was sung by a group of young people—later revealed as Hitler Youth—in a seemingly peaceful beer garden. (A couple of the eliminated numbers, incidentally, could still be heard as background music—with lyrics in German—over a radio.)

The characters of Ernst Ludwig and Herr Schultz were cut, and Fraulein Schneider was reduced to a bit. The American writer now became an Englishman named Brian Roberts, effectively played by Michael York. Joel Grey, who'd won a Tony for his stage performance, repeated as the Emcee, and Sally Bowles was perfectly cast in the person of Liza Minnelli.

"There are three new songs in the film," said Fosse, "all performed in the cabaret. We used a very small orchestra—like eight to ten pieces. There is no big, stereophonic sound. . . . To make the music more authentic, more like the real period of Berlin in the thirties, we cut the number of instruments way down. So it sounds like the orchestra you see on the screen . . . the sound is amazingly thick, heavy and full."

Shot on location in Germany, *Cabaret* is closer in

Cabaret. Liza Minnelli and Joel Grey sing "The Money Song."

Cabaret. Fritz Wepper and Michael York.

Cabaret. Michael York, Liza Minnelli, and Helmut Griem.

spirit to Isherwood's original *Berlin Stories* than to either the stage musical or Van Druten's play. The love story between Sally and the writer is still the primary thrust, however Brian Roberts is bisexual in his tastes, and not only has an affair with Sally but also dallies with a wealthy German aristocrat, Max (Helmut Griem), who is also interested in the loose Miss Bowles. Certainly one of the script's most effective pieces of dialogue occurs when Brian and Sally are arguing about the millionaire. "Screw Maximilian!" shouts Brian. "I do," Sally retorts. A pause, and Brian replies, "So do I."

There are two other new characters in the picture: Natalia (Marisa Berenson), the daughter of a rich Jewish businessman, and Fritz (Fritz Wepper), a fortune hunter who falls in love with her and is forced to admit that he is a Jew so that they may marry.

By confining his musical passages rather than opening them up as is the popular custom in this film genre, Fosse choreographed an electrifying series of numbers that brought the cabaret sequences to life in all their disquieting impropriety. Most contained a garish array of half-naked chorines in black stockings, who surrounded Grey and Miss Minnelli as the pair warbled words of social commentary that reflected on the current state of the plot. (Grey's rendition of "Willkommen," for example, was cross-cut with Brian's arrival in Berlin.) "The Money Song" and "Mein Herr," performed with out the chorus, are prize musical sequences—perhaps among the best ever filmed.

"Everybody in *Cabaret* is very fine," said Roger Greenspun in the *New York Times,* "and meticulously chosen for type, down to the last weary transvestite and to the least of the bland, blond open-faced Nazis in the background. As for Miss Minnelli, she is sometimes wrong in the details of her role, but so magnificently right for the film as a whole that I should prefer not to imagine it without her."

Come Academy Award time, *Cabaret* was, unfortunately, up against *The Godfather.* The Allied Artists musical won eight statuettes—including best actress, supporting actor (Grey), and director. But when it came to announcing best picture, presenter Clint Eastwood opened the envelope and called off the name of the gangster film classic. It had also won best actor (Marlon Brando) and best screenplay. *Cabaret* creators had to console themselves with the knowledge that their film grossed over twenty million dollars in domestic film rentals.

Since 1972, there has been no mad stampede on the part of filmmakers to take stage musicals and shoot them as dramas-with-music rather than in their original integrated form. But if such a trend should ever develop, the producers would be wise to use *Cabaret* as their point of reference.

23
Man of La Mancha
(1972)

"**Perhaps** *Man of La Mancha* should never have been made into a film," reflects Arthur Hiller, who both produced and directed the 1972 screen adaptation of the play. "It's one of those special pieces of theater that works brilliantly on stage where reality is easily suspended, but suffers when placed before the cold eye of the movie camera."

Dale Wasserman's hit musical was inspired by a quotation from Miguel de Unamuno, "Only he who attempts the absurd is capable of achieving the impossible." It tells the story of Miguel de Cervantes—soldier, playwright, actor, tax collector, frequent jailbird—a failure for most of his life. Tossed into a Seville dungeon to await trial by the Inquisition, he is tried by his fellow inmates in a kangaroo court and, as his defense, recites his story of Don Quixote, the demented Spanish gentleman who fancies himself a knight of old.

To capture the tough and tender spirit of Cervantes, and to merge his identity with that of his classic character, Wasserman utilized a "play within the play" device here. Cervantes enacts the role of Quixote and the other prisoners play supporting parts.

The fantasy sequence has the would-be knight, accompanied by his faithful Sancho Panza, tilting with a windmill, visiting an inn, which he takes for a castle, imagining that a muleteers' whore is his refined lady love, and battling a whole slew of ruffians with his crooked lance.

Though the play concludes with Quixote's death and the summoning of Cervantes by the Inquisition, audiences walk away feeling ennobled. The message of Wasserman/Cervantes has been driven home: "No matter what the odds, man must continue to strive upward . . . to dream the impossible dream."

Wasserman originally wrote this work as a ninety minute non-musical television program, produced in 1959 by David Susskind for the CBS "DuPont Show of the Month." Its title then was *I, Don Quixote* and the star was Lee J. Cobb.

Not satisfied that his material had achieved its full potential in its video production, Wasserman reworked it, striving for more of a stylization than had been presented on the tube. Now a potential stage play, it was optioned but never produced. Wasser man was relieved. He still felt that some key element was missing, yet he wasn't sure what that element was.

Director Albert Marre *(Milk and Honey)* read the piece and phoned Wasserman with the answer to the riddle. The play should be turned into a musical. The playwright agreed. Mitch Leigh was chosen to compose the score for *Man of La Mancha* and Joe Darion to write lyrics.

Financial backers and producers for this "too

Man of La Mancha. Richard Kiley played Don Quixote/
Miguel Cervantes in the original stage production.

Man of La Mancha. **James Coco and Peter O'Toole.**

special, too intellectual" entertainment were not easy i:o come by. After an initial production at the Goodspeed Opera House in Connecticut, *La Mancha* snuck into New York with little publicity, a miserable advance ticket sale, and no recording company interested in doing an original cast album. It opened on November 22, 1965 at the ANTA Washington Square Theater in Greenwich Village. Producers were Albert W. Selden and Hal James. Marre directed, Jack Cole choreographed, and the ingenious single island-stage setting was created by Howard Bay. The cast included Richard Kiley as Don Quixote/Cervantes, Irving Jacobson as Sancho, Joan Diener as Aldonza, the trollop, and Ray Middleton as the Innkeeper.

Presented on a darkened stage without intermission, the dramatic musical (budgeted at $200,000) made use of various props to change the setting to the demands of the script which, to a great extent, took place "in the imagination of Miguel de Cervantes."

The critical reviews were ecstatic and took the New York theater crowd by surprise. Howard Taubman in the *New York Times* said, "Audacious in conception and tasteful in execution . . . there are charm, gallantry and delicacy of spirit in the rein carnation of Don Quixote"; and Emory Lewis of *Cue Magazine* called it "the best musical of the season."

Man of La Mancha was sold out for years, playing a total of 2328 performances. It took the New York Drama Critics award as best musical, and won six Tonys, including best musical and best musical star (Kiley). Three years after it had opened, the play had grossed over fifty million dollars from performances and record sales. It also gave the world one of its most inspiring and often-recorded songs in the field of popular music—"The Impossible Dream."

Anthony Quinn was among the first to bid for the *La Mancha* movie rights. He wanted to play Quixote to Cantinflas' Sancho. Later, Richard Burton was often discussed to recreate the Kiley role on film. When the picture rights were finally sold in 1968, they went to United Artists for three million dollars plus a hefty percentage of the gross profits, along with the proviso that filming could not commence as long as the play was still on the boards in any first-class theaters. Wags projected that that wouldn't be until 1971.

Marre was the first director on the movie. He scouted some locations and did a few screen tests, but was subsequently replaced by Peter Glenville. According to insiders, Glenville wanted to make Cervantes' *Don Quixote,* not Wasserman's play. Since Wasserman had already been contracted to write the screenplay, this director didn't last too long either.

Arthur Hiller, whose previous credits included *Love Story* and *The Hospital,* joined the project as producer/director late in the planning. Already cast by Glenville were the three principal roles—Peter O'Toole as Cervantes/Quixote, Sophia Loren as Aldonza, and James Coco as Sancho. Set designs by Luciano Damiani had also been completed. Hiller: "I was agreeable to the casting or I wouldn't have joined the venture. I chose Harry Andrews to play the Innkeeper. The sets caused me no particular problems and I altered them to suit my needs."

Interviewed on the film's location in Rome, Hiller discussed the difficulties he was encountering: "The essential problem of transferring this play to film is that the play takes place in the mind's eye. We have to figure out how to establish the prison-reality, the fantasy, and make the transitions between them. We need a mind's-eye effect—not a radical stylization as in *Fellini Satyricon.* Just an awareness that the action takes place in the mind of Miguel de Cervantes."

Unfortunately, Hiller decided to play up the realistic aspects in *La Mancha* and thereby eliminated most of the conventions so vital to successful musical theater. Dance numbers were choreographed so they would not look like dance numbers. Thus the muleteers rape of Aldonza, tastefully handled on stage in Jack Cole's choreography, became a disturbingly brutal sequence in the screen enactment. Wasserman's script, including the Don Quixote fantasy sections, was played for drama and as a result any comedy or light touches inherent in the lines were virtually lost. Settings, both interior and exterior, emphasized the poverty of the period. As Peter O'Toole put it: "They're the most incredibly depressing sets that have surely ever existed."

Another error in judgment occured when it came

Man of La Mancha. James Coco and Peter O'Toole.

Man of La Mancha. Sophia Loren and player.

Man of La Mancha. Julie Gregg, Ian Richardson, and Rosalie Crutchley sing "I'm Only Thinking of Him."

to handling the musical aspects of the nine million dollar picture. *La Mancha* has a truly stirring score that requires excellent singing voices to put it across properly. Neither O'Toole or Miss Loren possesses such vocal gifts, yet the lady did her own numbers—with mixed results. Songs like "Aldonza," the success of which were dependent on good acting rather than voice, still worked well, but the more melodic ones fell flat. O'Toole did his own singing while the film was being shot—to get the proper intonation and mood into each piece—then allowed himself to be over-dubbed later. In this post-production process, the emphasis was put on matching O'Toole's vocal characteristics, rather than achieving proper resonance. The final result was that songs written to be *sung* by the leading players were *acted*—a valid approach considering Hiller's realistic conception of the material, but certainly disappointing to viewers who'd hoped for a proper delivery of the magnificent Darion/Leigh score.

(Some audience members would comment that Hiller's serious film might have worked better if the words and music had been eliminated altogether. That way, he would have at least had a good dramatic film, rather than a musical in which the songs almost seemed intrusive to the mood.)

La Mancha did have its good points—particularly in the *dramatic* performances of its principals and the singing voices of its supporting cast. O'Toole played Cervantes with considerable restraint, then—in spite of the picture's overall tone—gave a delightfully looney accounting of Quixote. Miss Loren was fiery, yet vulnerable, as Aldonza. Coco, Andrews, Ian Richardson (as the Padre), and Julie

Man of La Mancha. **James Coco, Peter O'Toole, and Sophia Loren.**

Man of La Mancha. **Peter O'Toole.**

Gregg (as Quixote's niece) also made memorable contributions.

Richard Cuskelly, writing in the *Los Angeles Herald-Examiner*, put the film's basic problem quite simply when he said: "Director Hiller has failed to arouse the emotions or make the magic work. His direction is intelligent and purposeful, but Don Quixote's illusions have eluded him."

Other reviewers were not so kind as Cuskelly. Many found the pace sluggish, and *Newsweek* even tagged it as an "animated classic comic of the great Cervantes novel." The picture flopped miserably at the box office.

Is *Man of La Mancha* one of those plays that—because of its basic unreality—defies adaptation to the screen? Or could another director—more experienced than Hiller in musical techniques—have licked the script's inherent problems?

Had Hiller dropped his realistic approach in the Quixote sequences only—playing them more broadly and against less somber settings-would Wasserman's stage magic have come through? Staging his exterior scenes against a cyclorama, for example, rather than out-of-doors, might have made an interesting difference.

As Hiller himself has suggested, would an overall concept of more fantasy have made the movie work?

Whatever the answers to these academic questions might be, Hiller's *Man of La Mancha* certainly proves that reality can be the book musical's deadliest enemy, and that trying to mix the two techniques doesn't always work.

24
Jesus Christ Superstar
(1973)

Rock musicals have become an exciting part of the theater scene in recent years. Shows like *Godspell, Hair,* and *Tommy* have stunned critics, thrilled audiences, and made millions of dollars for their creators. *Grease,* with its score inspired by the rock sound of the fifties, has become the all-time box office champ of musical films.

One of the most successful entries in this musical genre was the thoughtful, deeply concerned *Jesus Christ Superstar*—a project that began life as a song, "Superstar," then developed into a two-disc 87-minute recording. Available at first only in Great Britain and achieving only limited success, this rock opera album, produced, written and performed by Andrew Lloyd-Webber and Tim Rice, was brought to the attention of film director Norman Jewison, who was then in Yugoslavia shooting *Fiddler on the Roof.* "The morning after listening to it all night," reported Jewison, "I was so excited I wired the head of Universal Studios who held the rights [in conjunction with the Robert Stigwood Organization], and for the first time in my life committed myself impulsively to a film."

Universal had made a 50/50 financing deal with Stigwood, who'd been involved with the property since its inception. He'd been staging live concerts of the music and controlled it. With Universal's help, he would now develop *Jesus Christ Superstar* into a stage production and later into a motion picture. Lloyd-Webber and Rice would devise the legit version, budgeted at around seven hundred thousand dollars.

The album, distributed in the United States by Decca, suddenly began to take off with the record buying public. It became a best-seller for over a year. Then the stage production opened in October of 1971 at the Mark Hellinger Theater. Tom O'Horgan, who'd also done *Hair,* directed. The show received a divided reaction from critics. Reviewers either loved it or hated it. "I felt it wasn't worthy of the furore, enthusiasm and ire that it has aroused," was the reaction of Richard Watts in the *New York Post.* Douglas Watt of the *Daily News,* on the other hand, called it "a triumph."

Jesus Christ Superstar was the story of Christ's last seven days, told provocatively through pop music and song. Treating the subject matter with complete reverence, but with a twentieth century frame of reference, the director filled his bare, raked stage with color, action, and special effects. Dazzling set pieces and even actors were raised and lowered from the flies. Characters in the play carried hand mikes and made the most bizarre entrances imaginable. King Herod, for example, emerged

Jesus Christ Superstar. **Ted Neeley as Jesus.**

from a green monster head, lying on a pink shell.

Judas, the strongest, most demanding role in the production, was played brilliantly by a black baritone actor, the then unknown Ben Vereen. Jeff Fenholt, a tenor, was Jesus, while Yvonne Elliman—who with Barry Dennen as Pontius Pilate were the only cast members from the original record—played Mary Magdalene. She was quite moving in her rendition of "I Don't Know How to Love Him." As King Herod, Paul Ainsley stopped the show briefly when he sang his challenge to Jesus, "Turn my water into wine . . . ."

The marvelous Lloyd-Webber/Rice rock score and O'Horgan's staging combined to make an evening of probing theater that fascinated and entrapped audiences—especially the younger set. What the critics said was of no importance. The musical sold out for weeks in advance. Soon touring

Jesus Christ Superstar. **Carl Anderson as Judas.**

companies and regional productions of the show were authorized by Stigwood in the United States and various other countries. Ninety days after the New York opening, the promoter/producer predicted the traveling and Broadway companies com bined would gross about twenty million dollars in their first year.

When Norman Jewison, who would co-produce the screen adaptation with Stigwood, begn preparing his film project in earnest, he made it clear that he definitely didn't want to do anything like the New York stage version. "Rather too plastic, I fear, for my taste," he told the press.

"The one thing I knew for sure I didn't want was a *King of Kings* job. I've seen Pasolini's *The Gospel According to St. Matthew* at least eight times; it's so spare and simple and close to the Bible—and that's what I had in the back of my mind .

"The first scenario that Andrew Lloyd-Webber and Tim Rice came up with was pure *Kings of Kings* with all the trappings: a cast of thousands, you know. They had this very modern concept for the music but when it came to the visuals they lapsed right back to sheer Hollywood thirties. I went to work on it myself with Melvyn Bragg."

Working with a $3.5 million budget, Jewison decided to film his production on location in Israel, primarily in remote, dramatic spots. "I saw it played against wide expanses of hot dry sand and ruins," he said. "We wanted to stay far away from any religious colored picture postcard aspect in choosing locations."

His cast was made up entirely of unknowns. "It's

Jesus Christ Superstar. **Ted Neeley as Jesus and Carl Anderson as Judas.**

Jesus Christ Superstar. Yvonne Elliman as Mary Magdalene.

impossible to get the kind of performance an opera needs from people who can't sing," he said to the *Hollywood Reporter*. "But the problem was to find people to act the parts, to give the feeling of talking, not singing.

"It was a hot, arduous location but they learned quickly. Their tremendous energy and intensity is evident through the film. The lack of experience resulted in unusually honest performances, almost as if they were unaware of the camera. They all moved well and were used to music. I found that they responded if I carefully staged their movements and then moved the camera around them."

Barry Dennen, who'd been in *Fiddler* for Jewison and had brought the original *Jesus Christ Superstar* album to the director's attention, and Yvonne Elliman were the lone members of the Broadway company to do the movie. For the angry Judas, Jewison cast Carl Anderson; Jesus was Texas-born

Jesus Christ Superstar. The followers of Jesus dance in the desert.

Ted Neeley, and Joshua Mostel (Zero's son) played Herod.

The picture begins with the cast—all young people—piling off a dusty bus somewhere in the desert. "Maybe they're a traveling roadshow company," says the director, but you never learn for sure." The group sets up its props, then immediately does the first number. The familiar biblical story proceeds from there, with the ideological conflict between Jesus and Judas serving as the backbone. Christ is costumed in the traditional white robe, but his Apostles are garbed in what might best be described as "California mod." Modern props—sub-machine guns for the soldiers—are freely mixed with those in keeping with the story's period.

The 1973 Universal release was peppered with fanciful images: Judas being pursued by tanks across the desert, Caiaphas (Bob Bingham) playing his scenes on a skeletal structure reminiscent of a spider-web, the Last Supper staged as a picnic, and, finally, Christ being nailed to the cross by surfer types in hard hats.

Choreographer Ron Iscove designed a number of unusual dance patterns, including one in which the dancers appear in the desert before Christ and jump about like birds of sprung steel. The climactic number, "Superstar," is staged as one glorious rock light show.

At the picture's conclusion, the players return silently to the bus. The actor who portrayed Jesus is absent.

Performances in the film were generally excellent, with Anderson and Miss Elliman taking full advantage of the two most interesting parts. Neeley was also effective in a role that is, historically, impossible to enact with complete confidence. Mostel, playing

Jesus Christ Superstar. **Joshua Mostel (c) as King Herod.**

Herod on the swishy side, was nevertheless amusing in one of the show's more delightful moments.

As with the stage production, critics and audiences were divided into two distinct camps. They either loved or hated the picture. There was no middle ground. *Hollywood Reporter* called it "one of the most exquisitely put together movies in quite some time. Sometimes gloss can be boring but not in this film." Conversely, *Newsweek* announced the film was "one of the true fiascos of modern cinema." In spite of this mixed *reaction, Jesus Christ Superstar* grossed over thirteen million dollars in domestic rentals.

A highly charged musical entertainment, Superstar reflected both man's despair with organized religion and his innate need for some sort of life-organizing faith beyond self. It was a film that could, depending on the individual viewer, either reopen or reinforce his personal beliefs about God and universe.

25

A Little Night Music

(1977)

Even when the original stage director tackles the movie version of a musical, there's no guarantee the results will be satisfying.

Harold Prince and Stephen Sondheim had wanted to do a musical with a score made up entirely of waltzes ever since 1957, when they worked together on *West Side Story*. It was well over a decade before they proceeded seriously with their plan and, with the aid of playwright Hugh Wheeler, began looking for a suitable property. Sondheim remembered a movie he'd seen years before—Ingmar Bergman's *Smiles of a Summer Night* (1955)—and suggested that this comedy might be just what they were looking for.

With Bergman's approval, Wheeler wrote the book while Sondheim tackled music and lyrics. Prince both produced and directed the project, choreographed by Patricia Birch. The show's new title was *A Little Night Music*. The cast of the original production, which opened at the Shubert Theater in February of 1973, included Glynis Johns, Len Cariou, Hermione Gingold, and Laurence Guittard.

Reminiscent of Chekhov, the plot of this stylish operetta is set in Sweden at the turn of the century. Desiree Armfeldt (Miss Johns) is an aging traveling actress who once had an affair with lawyer Fredrik Egerman (Cariou). Egerman, who has recently wed an eighteen-year-old girl—still a virgin after eleven months of marriage—is unaware that he is the father of Desiree's daughter, Fredrika. The child lives on a country estate with her wealthy grandmother (Gingold), once the mistress of the king of Belgium.

After renewing her relationship one night with the sexually-frustrated Egerman, Desiree informs the lawyer that she has a lover, Count Malcolm (Guittard). Enter the jealous count, nearly catching the couple in a compromising position. It is the actress' quick thinking that prevents Egerman from facing Malcolm in a duel.

Desiree wants to settle down, and she has her eye on Egerman. She asks her mother to invite the lawyer, his wife and grown son, Henrik, to spend a weekend at the country estate. Hearing of this, an infuriated Malcolm decides to drop in on the gathering. His wife, Charlotte, hoping to win back her husband's affections, accompanies him.

Several tense, amorous, and revealing confrontations occur at the estate that warm evening. The bittersweet play concludes with the count reconciling with Charlotte, Henrik, who'd always been smitten with his father's wife, running off with the

A Little Night Music. Len Cariou, Elizabeth Taylor, Laurence Guittard and Diana Rigg.

young girl, and a shaken Egerman deciding to stay with Desiree.

A Little Night Music was a beautifully designed (by Boris Aronson), delicately staged production. It featured a strolling "Greek" chorus of five singers—dressed in evening clothes—who from time to time, come on stage and make their musical comments. The waltz-like score was pleasant, if not hummable-something one would expect in a play based on one of Bergman's films. (One song—"Send in the Clowns"—has become a popular standard.) Performances were all first-rate.

Douglas Watt in the *Daily News* said: "Everywhere, in fact, *A Little Night Music* reveals the work of superior theatrical craftsmanship. But stunning as it is to gaze upon and clever as its score is, with its use of trio and ensemble singing, it remains too literary and precious a work to stir the emotions."

A Little Night Music. Laurence Guittard and Diana Rigg.

The show won the Tony award for best musical of the season, and played a total of 601 performances. Jean Simmons, incidentally, starred in the national company version of *Music*, essaying the part created by Miss Johns.

With its chamber opera design, *A Little Night Music* was an unusual show that appealed to a "special," more. discriminating audience. "The audience has to make some eff ort, which people begrudgingly do," Hugh Wheeler told the *Los Angeles Times*. "Secretly, they'll never like it as much as something easier, but you have to decide whether to fawn on your audience or try to drag them a little bit in your direction."

This being the case, it's not surprising that Hollywood did not rush in to grab the play's movie rights. It was Elliott Kastner, a London-Based independent U.S. producer, who arranged to bring Night Music to the screen. He'd seen the show four times, obviously loved it, and, knowing that the Austrian government was seeking properties with which it could enter international film production, made a deal with that country's Sascha Wien Films for financing. Part of the picture's $7 .5 million budget also came from West German investors.

Harold Prince was retained to direct the production which, because of the financial aspects of the project, would now be reset in Austria. Hugh Wheeler did the screenplay, Sondheim made the proper alterations in his music, and Patricia Birch again choreographed.

Transfering the Sondheim/Wheeler work to the screen required a number of adjustments on the part

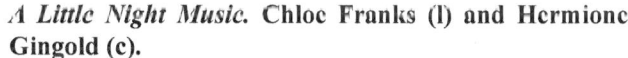

A Little Night Music. **Chloe Franks (l) and Hermione Gingold (c).**

A Little Night Music. **Laurence Guittard and Diana Rigg.**

A Little Night Music. **Len Cariou and Laurence Guittard.**

of the creators—particularly in uplifting the general tone. Said Prince: "On the stage it was what we called our Chekhov musical—with long pauses for sighs and regrets, and time to muse. The Chekhov didn't seem appropriate in the film. The characters in the film are more accessible, the emotions are more Middle European and recognizable. I think the emotions in the play and in the Bergman movie were dictated by that summer night in Sweden—it was exotic, unfamiliar.

"We got the idea of making the movie more of a pastry, and for that reason the switch to Austria was perfectly natural."

The move of location, in itself, dictated some changes. One number "The Sun Won't Set," with lyrics about Swedish summer days, was given new words and titled, "Love Is a Lecture." Several other songs ("The Glamorous Life," "Liaisons," "The Miller's Son") were dropped altogether.

Gone, too, from the movie was the strolling chorus. Prince began and ended his picture inside Vienna's 175-year-old Theater an der Wien. Under the film's credits, a turn of the century audience take their seats, then the curtain goes up and we see the show's principals on a woodland set, singing and dancing to "Love Is a Lecture." This setting fades, and the players dance off into an actual exterior. The play begins.

"In this way," says Prince, "the moviegoers will be led by degrees into accepting the relative artificiality of people singing on the screen in closeup. Using the theater also underlines the metaphor which is the point of the piece—that people wear masks and play roles that get them into trouble."

Casting of *A Little Night Music* was a mixture of original cast members, and Hollywood "box office names." Retained from the stage were Hermione Gingold and Laurence Guittard, while Elizabeth Taylor was signed to star as Desiree, Diana Rigg as Charlotte, and Robert Stephens as Egerman. A three-week rehearsal period had just gotten under way when Stephens was "fired" by Prince for "private" reasons. Insiders have credited the dismissal to a personality conflict between the actor and Miss Taylor. Len Cariou was rushed in to replace Stephens, who reportedly had himself been a replacement for Peter Finch.

Once the movie had been completed, it sat on the shelf for months, seemingly unable to attract a major distributor. Some rumors said the film would be sold directly to television. Others claimed it was an "artistic turkey." Virtually the entire industry was aware that, even if *Night Music* was a gem, it was still not the sort of film a mass audience would readily buy. The book was just too literate for the average taste.

In any event, Roger Corman's New World Pictures finally picked up the picture for distribution, and booked it into its initial playdates late in 1977. Business was spotty.

The stage version of A Little Night Music had seemed to be almost a steady stream of enchanting melodies—due to a great extent to the strolling quintet of singers, who moved in and out of the action and in general kept things interesting during

A Little Night Music. Elizabeth Taylor.

any slow moments. This chorus was one of the most exciting elements in the original staging. Without it in the movie, the book was robbed of much of its charm, and was forced to include long stretches of dramatic action without any music. Under the direction of Prince, whose previous film experience was limited, *A Little Night Music* moved far too slowly. Sondheim's music was the best thing about the production, yet his songs were spaced far apart. One of the more memorable numbers was "A Weekend in the Country," which included cross talking lyrics for four voices—a reminder of the musicals from Hollywood's Golden Era.

The film's technical aspects were superb, with special credit due Arthur Ibbetson's photography and the art direction of Herta Pischinger. A 264-year-old chateau near Vienna, the Schloss Schoenborn, served as the movie's country estate.

For the most part, performances were excellent, with best notices going to Miss Rigg, Cariou, Guittard, and Lesley-Ann Down, who played the virgin wife. Though slightly overweight, Elizabeth Taylor was acceptable as Desiree, although she seemed insecure with her singing voice. Prince might have been wiser to utilize Jean Simmons in the part. She still has a fair box office following, though she lacks the magic of the Taylor name. Based on her stage performance, Simmons would have given a more satisfactory accounting of the part.

Ron Pennington, writing in the *Hollywood Reporter,* said of the film: "In spite of the flaws, however, it is not really a bad movie. It's just a disappointing one, especially in consideration of Prince's innovative stage work." The film did win the 1977 Academy Award for best musical scoring—adaptation.

Little Night Music. **Laurence Guittard, Elizabeth Taylor, Christopher Guard, Lesley-Ann Down, Len Cariou, Diana Rigg, and Hermione Gingold.**

Perhaps if Prince had stayed closer to what he'd had on the stage and tried to find a way of translating the elements of that production into filmic terms, he would have wound up with a much better picture.

A Musical Gallery

The stills on the following pages represent only a portion of the many motion pictures that have had their basis in the musical theater.

Animal Crackers (1930). Zeppo Marx, Groucho Marx, Margaret Dumont, and players. Harry Ruby and Bert Kalmar wrote the songs for this 1928 Marx Brothers stage romp. The film was a virtual record of the stage production.

Anything Goes (1936). Ida Lupino and Bing Crosby, who also starred in the 1956 remake, played in this loose adaptation of the 1934 Cole Porter musical. Ethel Merman, star of the stage version, was also in the movie.

The Boys from Syracuse (1940). Eric Blore, Alan Mowbray, Allan Jones, and Joe Penner starred in Universal's rendition of Rodgers and Hart's 1938 musical, based on Shakespeare's *The Comedy of Errors*.

Babes in Arms (1939). Mickey Rooney and Charles Winninger. Produced on Broadway in 1937, the Rodgers and Hart musical became a perfect vehicle for Rooney and Judy Garland.

Babes in Toyland (1934). Victor Herbert's 1903 musical extravaganza was adapted to the madcap talents of Laurel and Hardy.

The Band Wagon (1953). Cyd Charisse and Fred Astaire in one of Astaire's best films. It was originally a revue by Arthur Schwartz and Howard Dietz that debuted in New York in 1931.

Bells are Ringing (1960). Judy Holliday and Dean Martin sing "Just in Time" from the Comden/Green/Styne score of this 1956 stage musical. Miss Holliday was the show's original star.

Bitter Sweet (1940). Nelson Eddy and Jeanette Mac Donald starred in Metro's adaptation of the Noel Coward operetta, which played New York in 1929 for 151 performances.

The Boy Friend (1971). Sandy Wilson's clever 1954 spoof of the twenties became a dazzling and confusing musical extravaganza in the hands of director Ken Russell. The MGM release starred Twiggy in the role performed on stage by Julie Andrews.

Brigadoon (1955). Gene Kelly and Cyd Charisse do what they do best in this Vincente Minnelli/MGM musical. The Lerner/Loewe show debuted on Broadway in 1947.

Cabin in the Sky (1942). Vincente Minnelli directed Kenneth Spencer, Eddie "Rochester" Anderson, Lena Horne, and Rex Ingram in this MGM production. The 1940 stage show had a score by Vernon Duke, John Latouche, and Ted Fetter.

The Belle of New York (1952). Vera-Ellen and Fred Astaire perform "A Bride's Wedding Day Song." The original stage version, with songs by Gustave Kerker and Hugh Morton, debuted in 1897 and became the first American musical to play London's West End. Johnny Mercer and Harry Warren wrote the new film score. Charles Walters directed.

Call Me Madam (1953). Ethel Merman, seen here with Billy DeWolfe, Walter Slezak, and Steven Geray, reprised her original Broadway role in this film version of the 1950 Irving Berlin musical.

Call Me Mister (1951). Betty Grable was the star of the movie version of Harold Rome's 1946 musical about soldiers facing post-war demobilization problems.

Camelot (1967). Franco Nero, Richard Harris, and Vanessa Redgrave replaced stage stars Robert Goulet, Richard Burton, and Julie Andrews in this lush Warner Brothers release. The Lerner/Loewe show debuted in 1960.

Can Can (1960). Shirley MacLaine performs the infamous dance of the title in this Fox version of Cole Porter's 1953 musical. The film score interpolated songs from the composer's earlier shows.

A Connecticut Yankee (1948). Bing Crosby and Rhonda Fleming. The Rodgers and Hart score was dropped from this adaptation and a new one added. The stage version debuted in 1927.

The Desert Song (1943). Dennis Morgan. This was the second of three movie versions of the 1926 Sigmund Romberg/Otto Harbach/Oscar Hammerstein II/Frank Mandell operetta.

DuBarry Was a Lady (1943). Lucille Ball and Gene Kelly. MGM retained only two of the original Cole Porter songs for this adaptation. Ethel Merman starred in the 1939 Broadway version.

The Desert Song (1952). Kathryn Grayson and Gordon MacRae were the leads in the most recent movie version of this classic operetta.

Fanny (1961). Horst Buchholz and Leslie Caron. This adaptation of the 1954 musical was a fine film—even though the Harold Rome score was relegated to background music.

Flower Drum Song (1961). Nancy Kwan and Jack Soo co-starred in Ross Hunter's over-produced version of the 1958 Rodgers and Hammerstein musical.

Funny Face (1956). Kay Thompson and Fred Astaire. Produced on Broadway with Adele and Fred Astaire in 1927, this Gershwin musical finally came to the screen with a new story line and a doctored score.

Funny Girl (1968). Barbra Streisand played Fanny Brice in both the movie and stage versions of this Jule Styne/Bob Merrill musical biography, which debuted on Broadway in 1964.

A Funny Thing Happened on the Way to the Forum (1966). Zero Mostel and Jack Gilford. Both starred in the 1962 stage production, which featured a score by Stephen Sondheim.

Gigi (1958). Leslie Caron and Louis Jourdan. This Oscar-winning (Best Picture) musical, with a Lerner/Loewe score, was adapted to the Broadway stage in 1973.

Girl Crazy (1943). Mickey Rooney, Judy Garland, and Tommy Dorsey. The 1930 stage version featured a score by the Gershwins, including "I Got Rhythm." Roger Edens contributed some new material to the film score.

Godspell (1973). This rock musical, a youth-slanted reworking of the Gospel according to St. Matthew, began as an off-Broadway production in 1971. The score was by Stephen Schwartz. David Greene directed the film, featuring a cast of unknowns.

Good News (1947). Joan McCracken leads the rousing title song in this second film version of the 1927 De Sylva, Brown, and Henderson college musical, which played 557 performances. June Allyson and Peter Lawford starred in the film.

The Great Waltz (1938). (l to r) Luise Rainer, Fernand Gravet, and Miliza Korjus. Hammerstein supplied lyrics for this film about Johann Strauss and his music. A 1934 stage version had featured lyrics by Desmond Carter, and a book by Moss Hart.

Hair (1979). With Milos Forman as director, this rock musical classic was an exciting film, but still lacked the impact of the original 1968 stage version. Music was by Galt MacDermot; book and lyrics by James Rado and Jerome Ragni. John Savage starred in the movie.

Grease (1978). Teen idol John Travolta (c) starred in this hit rock musical that was set in the fifties. The show, by Jim Jacobs and Warren Casey, had opened in New York in 1972. A Paramount release, *Grease* has become the highest-grossing movie musical of all time.

Half a Sixpence (1967). Julia Foster and Tommy Steele, who starred in the 1963 Broadway production. Based on *Kipps* by H. G. Wells, the show featured a score by David Heneker.

Hit the Deck (1955). Debbie Reynolds was one of several MGM stars in this version of the 1927 Vincent Youmans/Clifford Grey/Leo Robin/Irving Caesar musical.

Irma la Douce (1963). Shirley MacLaine and Jack Lemmon. When director Billy Wilder filmed his version of the 1958 hit musical, he eliminated the score by Marguerite Monnot.

Jacques Brei is Alive and Well and Living in Paris (1975). Elly Stone sings "Carousel" in the American Film Theatre's adaptation of the show that ran nearly five years Off Broadway, starting in 1968. Composer Brei also appeared in this disjointed film.

Kismet (1955). Howard Keel, Dolores Gray, Ann Blyth, and Vic Damone in this Vincente Minnelli-directed adaptation of the 1953 stage production. Music was by Borodin, as adapted by Wright and Forrest.

Knickerbocker Holiday (1943). Nelson Eddy and Constance Dowling. The Kurt Weill/Maxwell Anderson stage musical debuted in 1938.

Lady Be Good (1941). Eleanor Powell starred in this weak MGM filming of the smash 1924 Gershwin musical, which had starred Fred and Adele Astaire. The movie's big number was an interpolation by Kern/Hammerstein—"The Last Time I Saw Paris."

Lady in the Dark (1943). Ray Milland and Ginger Rogers were the stars of this Paramount release, based on the 1941 Kurt Weill/Ira Gershwin/Moss Hart musical.

Li'l Abner (1959). Stubby Kaye, Bern Hoffman, and Leslie Parrish starred in this Al Capp comic strip brought to life. The Gene de Paul/Johnny Mercer stage musical played New York in 1956.

Little Nellie Kelly (1940). Arthur Shields, Judy Garland, and Douglas MacPhail. George M. Cohan wrote the 1922 stage version, but many of the tunes in the movie were by Roger Edens. Garland played a dual role in this MGM release.

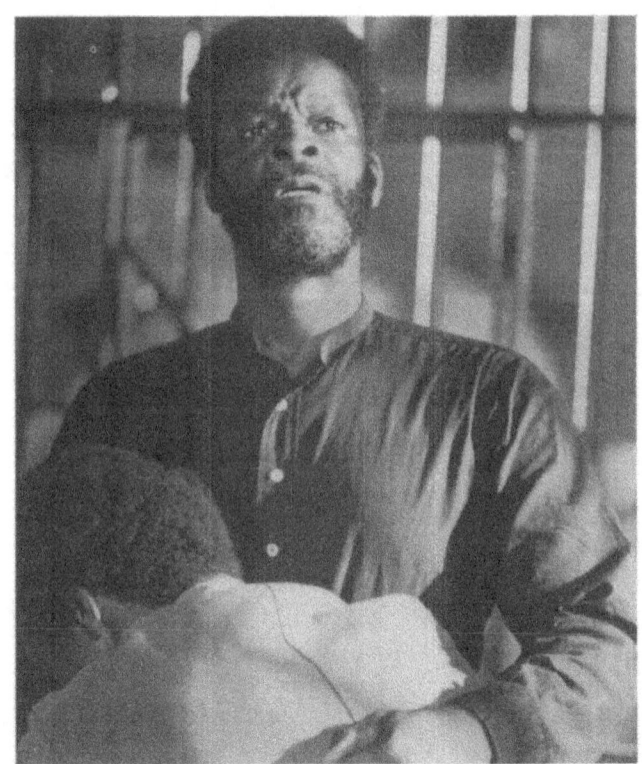

Lost in the Stars (1973). Brock Peters starred in this production for the American Film Theatre. Adapted from Alan Paton's novel, *Cry the Beloved Country*, the stage version, with music by Kurt Weill and book/lyrics by Maxwell Anderson, debuted in 1949.

Lovely to Look At (1952). Kathryn Grayson and Howard Keel starred in this remake of *Roberta*.

Mame (1974). Lucille Ball and Kirby Furlong. Angela Lansbury starred in the original 1966 stage musical, which featured a score by Jerry Herman.

Maytime (1937). Nelson Eddy and Jeanette MacDonald. Little of the original Sigmund Romberg score remained in this MGM adaptation. The operetta had debuted in 1917 with Peggy Wood and Charles Purcell in the leading roles.

The Merry Widow (1952). Lana Turner and Fernando Lamas were the stars of this dated remake of the popular Franz Lehar operetta, which had debuted in Vienna in 1905. Curtis Bernhardt directed the MGM production.

Oh! What a Lovely War (1969). Maggie Smith was one of the many stars who appeared in this disturbing antiwar satire that utilized interpolated songs and was directed by Richard Attenborough. The stage production played New York in 1963.

New Faces (1954). Virginia DeLuce, Ronny Graham, Robert Clary, and Charles Watts. Leonard Sillman's *New Faces of 1952* served as the basis of this Twentieth Century-Fox release.

On a Clear Day You Can See Forever (1969). Barbra Streisand was the star and Jack Nicholson had a minor role in this Vincente Minnelli-directed adaptation of the 1965 Burton Lane/Alan Jay Lerner show about ESP.

One Touch of Venus (1948). Robert Walker and Ava Gardner starred in this version of the 1943 stage production, which had headlined Mary Martin. The Universal release dropped much of the Kurt Weill score.

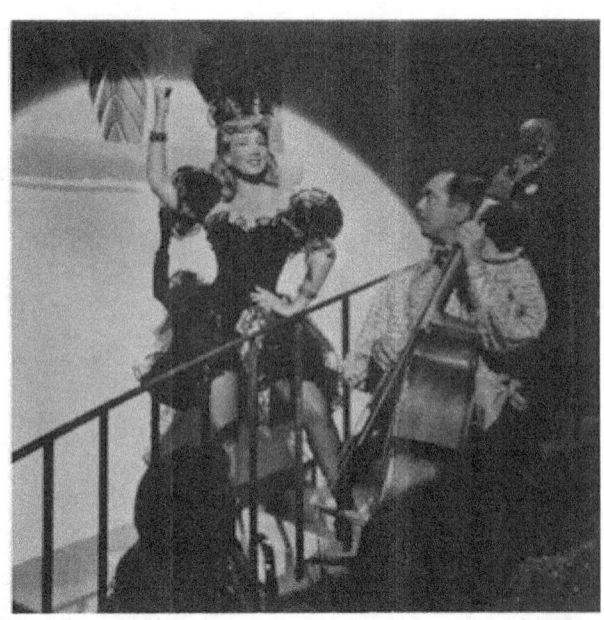

Panama Hattie (1940). Ann Sothern, inheriting the role created by Ethel Merman, sings Cole Porter's "I've Still Got My Health." Lena Horne and Red Skelton co-starred in the movie. The stage version had opened in 1940 for a 501 performance run.

The Pajama Game (1957). Doris Day leads the "7½ Cents" number in the hit Richard Alder/Jerry Ross show that played New York in 1954 with Janis Paige as its star.

Rio Rita (1942). The 1927 Harry Tierney/Joseph McCarthy musical became a framework for the antics of Abbott and Costello. Kathryn Grayson co-starred.

Roberta (1934). Fred Astaire and Ginger Rogers starred in this first version of the 1933 Jerome Kern musical.

Seven Brides for Seven Brothers (1954). Howard Keel and Jane Powell reprised the leading roles when this classic MGM musical—with a revised score—was adapted to the stage in 1978. The show closed before it reached Broadway.

Rosalie (1937). Ray Bolger and Eleanor Powell starred in the MGM movie, which featured a score by Cole Porter. The 1928 operetta had had its music written by Gershwin and Romberg. Marilyn Miller headlined the stage version.

1776 (1972). Howard da Silva played Ben Franklin on both stage and screen in this Tony award-winning musical, which played Broadway starting in 1969 for 1,217 performances. Music and lyrics were by Sherman Edwards.

Show Boat (1951). Howard Keel and Kathryn Grayson were the stars of MGM's lavish Technicolor remake of the Kern/Hammerstein musical.

Show Boat (1929). Joseph Schildkraut and Otis Harlan. This was the first movie version of America's classic musical, adapted from the novel by Edna Ferber.

Silk Stockings (1957). Fred Astaire and Cyd Charisse starred in this adaptation of the 1955 Cole Porter stage success, which in turn was based on the Garbo vehicle, *Ninotchka*.

Song of Norway (1970). Florence Henderson and Edward G. Robinson in this fictionalized biography of composer Edvard Grieg. Robert Wright and Chet Forrest adapted the Grieg music into the operetta, which played Broadway in 1944.

South Pacific (1958). Mitzi Gaynor and Rosanno Brazzi stood in for the original stars, Mary Martin and Ezio Pinza, in this classic Rodgers and Hammerstein musical, which debuted in 1949.

The Student Prince (1954). Edmund Purdom (with an assist from the singing voice of Mario Lanza) and Ann Blyth starred in this filming of the 1924 Romberg operetta. The MGM CinemaScope production was directed by Richard Thorpe.

Strike Up the Band (1940). This Judy Garland/Mickey Rooney vehicle utilized the title song and little else from the 1930 Gershwin musical. The bulk of the movie score was by Roger Edens and Arthur Freed.

Sunny (1930). Marilyn Miller, repeating her Broadway role, and Lawrence Gray. The 1925 hit had book and lyrics by Otto Harbach and Oscar Hammerstein II; music by Jerome Kern. It was the first Kern/Hammerstein musical, and introduced the song, "Who?"

Sweet Charity (1969). Paula Kelly, Shirley MacLaine, and Chita Rivera cavort under Bob Fosse's brilliant direction in the "There's Gotta Be Something Better Than This" number. The Cy Coleman/Dorothy Fields musical, starring Gwen Verdon, opened on Broadway in 1966.

Sweethearts (1938). Reginald Gardiner, Jeanette MacDonald, and Nelson Eddy. A 1913 Victor Herbert operetta was the genesis of this Technicolor picture, which featured a totally revamped storyline.

Tea for Two (1950). Doris Day and S. Z. Sakall in the second movie version of the Vincent Youmans/Irving Caesar/Otto Harbach musical, *No, No Nanette*, which had its stage debut in 1925. Miss Day's leading man was Gordon MacRae.

This is the Army (1943). Ronald Reagan, George Murphy, and Alan Hale helped supply a story framework when Warner Brothers filmed Irving Berlin's famous 1942 fund-raising soldier show. Champ Joe Louis, Kate Smith, and Berlin himself made special appearances in the picture.

Tommy (1975). Roger Daltry starred in Ken Russell's movie version of the rock opera by Peter Townshend, which began as a 1969 album by The Who and later was adapted to the concert and legitimate stages. Ann-Margret, Oliver Reed, and Elton John were also in the cast.

The Unsinkable Molly Brown (1960). MGM cast Harve Presnell and Debbie Reynolds in its adaptation of Meredith Willson's smash musical. The 1960 stage production starred Tammy Grimes.

The Vagabond King (1956). Kathryn Grayson and Oreste Kirkop starred in this dreadful remake of the Friml operetta. Michael Curtiz was director.

The Wiz (1978). This Black version of *The Wizard of Oz* featured many dazzling production numbers, and had a cast headed by Diana Ross. Charlie Small wrote the score for the 1975 stage success.

Where's Charley? (1952). Ray Bolger repeated his stage success in this Warner Brothers musical version of Brandon Thomas' *Charley's Aunt*. The show, boasting a Frank Loesser score, played New York in 1948. The most memorable of the tunes was "Once in Love with Amy," which became Bolger's theme song.

Index of Titles

All About Eve, 57
All That Money Can Buy, 70
American in Paris, An, 9, 18, 31
Anchors Aweigh, 28
Animal Crackers, 13, 167
Annie, 23
Annie Get Your Gun, 33–37, 43, 47
Anything Goes, 16, 43, 168

Babes in Arms, 16, 168
Babes in Toyland, 169
Band Wagon, The, 18, 169
Beach Blanket Bingo, 100
Belle of New York, The, 172
Bells Are Ringing, 18, 170
Berlin Stories, 143, 147
Best Years of Our Lives, The, 60
Big Boy, 13
Bitter Sweet, 170
Boy Friend, The, 19, 103, 171
Boys From Syracuse, The, 168
Brigadoon, 18, 21, 101, 119, 138, 171
Bye Bye Birdie, 96–100

Cabaret, 22, 143–47
Cabin in the Sky, 18, 172
Calamity Jane, 22, 23
Call Me Madam, 39, 95, 173
Call Me Mister, 173
Camelot, 21, 22, 141, 174
Can Can, 17, 174
Carmen Jones, 18, 19, 76, 77
Carousel, 61–65, 107
Cat Ballou, 142
Charley's Aunt, 199
Chicago, 73
Chorus Line, A, 23

Christmas Carol, A, 125
Collector, The, 110
Comedy of Errors, The, 168
Company, 23
Connecticut Yankee, A, 16, 175
Cover Girl, 68
Cry the Beloved Country, 186

Damn Yankees, 70–73, 103
Desert Song, The, 13, 17, 175, 176
Devil and Daniel Webster, The, 70
Don Quixote, 151
Dr. Dolittle, 137
DuBarry Was a Lady, 176

Easter Parade, 9

Fancy Free, 27
Fanny, 18, 176
Fantasticks, The, 23
Fellini Satyricon, 151
Fiddler on the Roof, 21, 22, 105, 155, 158
Finian's Rainbow, 119–24
Fiorello, 23
Flower Drum Song, 21, 177
Flying Down to Rio, 13
Follies, 23
Footlight Parade, 13
Forty-Second Street, 13
From Here to Eternity, 51
Funny Face, 177
Funny Girl, 16, 19, 20, 91, 133, 178
Funny Thing Happened on the Way to the Forum, A, 19, 178

Gay Divorce, The, 14, 43

Gay Divorcee, The, 14, 15
Gentlemen Prefer Blondes, 38, 42
Gigi, 9, 18, 22, 179
Girl Crazy, 16, 34, 179
Give 'Em Hell, Harry, 118
Godfather, The, 107, 124, 147
Godspell, 155, 180
Gone With The Wind, 107
Good News, 13, 180
Gospel According to St. Matthew, The, 157
Grease, 21, 22, 107, 155, 181
Great Waltz, The, 180
Green Grow the Lilacs, 49
Guys and Dolls, 55–60
Gypsy, 91, 95, 133

Hair, 23, 155, 181
Half a Sixpence, 115, 182
Hamlet, 115
Hammersmith is Out, 70
Hatlow, 115
Hello, Dolly!, 105, 131–37
High Noon, 51
Hit the Deck, 182
Hospital, The, 151
House of Flowers, 23
How to Succeed in Business Without Really Trying, 19, 20

I Am a Camera, 143
I Do, I Do, 23
I, Don Quixote, 148
Irma la Douce, 18, 183

Jacques Brei is Alive and Well and Living In Paris, 183
Jaws, 107
jazz Singer, The, 13
Jesus Christ Superstar, 22, 155–60
Jolson Story, The, 9

King and I, The, 20, 21, 64, 107
King of Kings, 157
Kipps, 182
Kismet, 18, 184
Kiss Me, Kate, 43–48
Knickerbocker Holiday, 184

Lady Be Good, 14, 184
Lady in the Dark, 185
Leave It to Me, 43
Li'l Abner, 185
Liliom, 61
Little Caesar, 93
Little Me, 23
Little Nellie Kelly, 186
Little Night Music, A, 161–66
Lost in the Stars, 18, 186
Love Story, 151
Lovely to Look At, 187

Mame, 21, 187
Man of La Mancha, 148–54
Mary Poppins, 105, 110, 130
Matchmaker, The, 131, 135
Maytime, 188
Meet Me in St. Louis, 18
Merry Widow, The, 17, 188
Mexican Hayride, 14, 15
Midnight Cowboy, 137
Milk and Honey, 148
Most Happy Fella, The, 23
Music Man, The, 86–90
My Fair Lady, 9, 21, 101–6, 109, 132, 133

New Faces, 189
New Faces of 1952, 189
New Girl in Town, 72

New Moon, 17
Ninotchka, 193
No, No Nanette, 197
No Strings, 23

Ohl What a Lovely War, 188
Oklahoma!, 18, 21, 27, 34, 43, 49–54, 60, 61, 62, 63, 75, 107
Old Curiosity Shop, The, 125
Oliver!, 9, 125–30
Oliver Twist, 113, 125
On a Clear Day You Can See Forever, 18, 189
One Touch of Venus, 190
On the Town, 27–32, 131, 137
Out of This World, 48

Paint Your Wagon, 138–42
Pajama Game, The, 21, 70, 72, 190
Pal joey, 17, 66–69
Panama Hattie, 190
Paramount on Parade, 13
Pickwick Papers, The, 125
Pipe Dream, 23
Pirate, The, 18
Porgy, 74
Porgy and Bess, 18, 49, 74–78
Professionals, The, 141
Pygmalion, 101

Random Harvest, 93
Redhead, 23, 73
Rio Rita, 13, 191
Roberta, 187, 191
Romeo and Juliet, 79
Rosalie, 14, 192
Rose Marie, 17

Sally, 13
Seven Brides for Seven Brothers, 9, 22, 192
1776, 21, 192
Show Boat, 1, 16, 47, 193
Show of Shows, The, 13

Silk Stockings, 193
Singin' in the Rain, 9, 31, 137
Smiles of a Summer Night, 161
Song of Norway, 194
Sound of Music, The, 9, 21, 107–12
South Pacific, 21, 34, 103, 107, 194
Star!, 137
Star Wars, 107
Stop the World—I Want to Get Off, 113–18
Strike Up the Band, 16, 195
Student Prince, The, 17, 195
Sunny, 13, 195
Sweet Charity, 21, 73, 196
Sweethearts, 196

Take Me Out to the Ball Game, 28
Tale of Two Cities, A, 125
Taming of the Shrew, The, 43
Tea for Two, 197
That's Entertainment, Part II, 124
This is the Army, 197
Three-Penny Opera, The, 143
Tommy, 22, 155, 198
Tora! Tora! Tora!, 137

Unsinkable Molly Brown, The, 20, 198

Vagabond King, The, 13, 14, 17, 199

West Side Story, 9, 27, 79–85, 112, 161
Where's Charley?, 199
Whoopee, 13
Who Was That Lady?, 72
Wildcat, 23
Wiz, The, 18, 22, 199
Wizard of Oz, The, 18, 21, 93, 130, 199
Words and Music, 28

Yankee Doodle Dandy, 9
Year the Yankees Lost the Pennant, The, 70